High-performance Packet Switching Architectures

Itamar Elhanany and Mounir Hamdi (Eds.)

High-performance Packet Switching Architectures

With 111 Figures

Springer

Itamar Elhanany, PhD
Electrical and Computer Engineering Department
The University of Tennessee at Knoxville
Knoxville, TN 37996-2100
USA

Mounir Hamdi, PhD
Department of Computer Science
Hong Kong University of Science and Technology
Clear Water Bay
Kowloon
Hong Kong

British Library Cataloguing in Publication Data
High-performance packet switching architectures
1.Packet switching (Data transmission)
I.Elhanany, Itamar II.Hamdi, Mounir
621.3'8216
ISBN-13: 9781846282737
ISBN-10: 184628273X

Library of Congress Control Number: 2006929602

ISBN-10: 1-84628-273-X e-ISBN 1-84628-274-8 Printed on acid-free paper
ISBN-13: 978-1-84628-273-7

Printed in Germany

9 8 7 6 5 4 3 2 1

Springer Science+Business Media
springer.com

Preface

Current estimates and measurements predict that Internet traffic will continue to grow for many years to come. Driving this growth is the fact that the Internet has moved from a convenience to a mission-critical platform for conducting and succeeding in business. In addition, the provision of advanced broadband services to end users will continue to cultivate and prolong this growth in the future. As a result, there is a great demand for gigabit/terabit routers and switches (IP routers, ATM switches, Ethernet switches) that knit together the constituent networks of the global Internet, creating the illusion of a unified whole. These switches/routers must not only have an aggregate capacity of gigabits/terabits coupled with forwarding rates of billions of packets per second, but they must also deal with nontrivial issues such as scheduling support for differentiated services, a wide variety of interface types, scalability in terms of capacity and port density, and backward compatibility with a wide range of legacy packet formats and routing protocols.

This edited book is a modest attempt to provide a comprehensive venue for advancing, analyzing, and debating the technologies required to address the above-mentioned challenges, such as scaling the Internet and improving its capabilities. In particular, this book is a collection of chapters covering a wide range of aspects pertaining to the design, analysis, and evolution of high-performance Internet switches and routers. Some of the topics include switching fabrics, network processors, optical packet switching and advanced protocol design. The authors of these chapters are some of the leading researchers in the field. As a result, it is our hope that this book will be perceived as a valuable resource to as many readers as possible including university professors and students, researchers from industry, and consultancy companies.

Acknowledgments

We would like to thank all the contributors of the book. Without their encouragement, enthusiasm and patience, this book would not have been possible. We would also like to thank Spinger for agreeing to publish this book. We wish to express our gratitude to Anthony Doyle (Engineering Editor) and Kate Brown (Engineering Editorial Assistant) for their careful consideration and helpful suggestions regarding the format and organization of the book. We also wish to thank Derek Rose for his enormous effort in helping us prepare this book.

Knoxville, USA, January 2006 Itamar Elhanany
Kowloon, Hong-Kong, January 2006 Mounir Hamdi

Contents

List of Contributors xi

1 Architectures of Internet Switches and Routers 1
Xin Li, Lotfi Mhamdi, Jing Liu, Konghong Pun, and Mounir Hamdi
1.1 Introduction 2
1.2 Bufferless Crossbar Switches 3
1.2.1 Introduction to Switch Fabrics 3
1.2.2 Output-queued Switches 4
1.2.3 Input-queued Switches 4
1.2.4 Scheduling Algorithms for VOQ Switches 5
1.2.5 Combined Input–Ouput-queued Switches 9
1.3 Buffered Crossbar Switches 12
1.3.1 Buffered Crossbar Switches Overview 12
1.3.2 The VOQ/BCS Architecture 13
1.4 Multi-stage Switching 19
1.4.1 Architecture Choice 19
1.4.2 The MSM Clos-network Architecture 20
1.4.3 The Bufferless Clos-network Architecture 23
1.5 Optical Packet Switching 27
1.5.1 Multi-rack Hybrid Opto-electronic Switch Architecture 27
1.5.2 Optical Fabrics 28
1.5.3 Reduced Rate Scheduling 30
1.5.4 Time Slot Assignment Approach 30
1.5.5 DOUBLE Algorithm 32
1.5.6 ADJUST Algorithm 32
1.6 Conclusion 34

2 Theoretical Performance of Input-queued Switches Using Lyapunov Methodology 39
Andrea Bianco, Paolo Giaccone, Emilio Leonardi, Marco Mellia, and Fabio Neri
2.1 Introduction 39
2.2 Theoretical Framework 41

2.2.1 Description of the Queueing System ... 41
2.2.2 Stability Definitions for a Queueing System ... 43
2.2.3 Lyapunov Methodology ... 44
2.2.4 Lyapunov Methodology to Bound Queue Sizes and Delays ... 47
2.2.5 Application to a Single Queue ... 48
2.2.6 Final Remarks ... 49
2.3 Performance of a Single Switch ... 50
2.3.1 Stability Region of Pure Input-queued Switches ... 51
2.3.2 Delay Bounds for Maximal Weight Matching ... 54
2.3.3 Stability Region of CIOQ with Speedup 2 ... 55
2.3.4 Scheduling Variable-size Packets ... 57
2.4 Networks of IQ Switches ... 58
2.4.1 Theoretical Performance ... 59
2.5 Conclusions ... 61

3 Adaptive Batched Scheduling for Packet Switching with Delays ... 65
Kevin Ross and Nicholas Bambos
3.1 Introduction ... 65
3.2 Switching Modes with Delays: A General Model ... 66
3.3 Batch Scheduling Algorithms ... 69
3.3.1 Fixed Batch Policies ... 70
3.3.2 Adaptive Batch Policies ... 72
3.3.3 The Simple-batch Static Schedule ... 73
3.4 An Interesting Application: Optical Networks ... 74
3.5 Throughput Maximization via Adaptive Batch Schedules ... 76
3.6 Summary ... 78

4 Geometry of Packet Switching: Maximal Throughput Cone Scheduling Algorithms ... 81
Kevin Ross and Nicholas Bambos
4.1 Introduction ... 81
4.2 Backlog Dynamics of Packet Switches ... 84
4.3 Switch Throughput and Rate Stability ... 86
4.4 Cone Algorithms for Packet Scheduling ... 88
4.4.1 Projective Cone Scheduling (PCS) ... 89
4.4.2 Relaxation, Generalizations, and Delayed PCS (D-PCS) ... 90
4.4.3 Argument Why PCS and D-PCS Maximize Throughput ... 92
4.4.4 Quality of Service and Load Balancing ... 93
4.5 Complexity in Cone Schedules – Scalable PCS Algorithms ... 95
4.5.1 Approximate PCS ... 95
4.5.2 Local PCS ... 95
4.6 Final Remarks ... 98

5 Fabric on a Chip: A Memory-management Perspective ... 101
Itamar Elhanany, Vahid Tabatabaee, and Brad Matthews
5.1 Introduction ... 101
5.1.1 Benefits of the Fabric-on-a-Chip Approach ... 102
5.2 Emulating an Output-queued Switch ... 103

5.3 Packet Placement Algorithm 105
5.3.1 Switch Architecture 105
5.3.2 Memory-management Algorithm and Related Resourses 106
5.3.3 Sufficiency Condition on the Number of Memories 109
5.4 Implementation Considerations 114
5.4.1 Logic Dataflow 114
5.4.2 FPGA Implementation Results 119
5.5 Conclusions 120

6 Packet Switch with Internally Buffered Crossbars 121
Zhen Guo, Roberto Rojas-Cessa, and Nirwan Ansari
6.1 Introduction to Packet Switches 121
6.2 Crossbar-based Switches 122
6.3 Internally Buffered Crossbars 124
6.4 Combined Input–Crosspoint Buffered (CICB) Crossbars 126
6.4.1 FIFO–CICO Switches 126
6.4.2 VOQ–CICB Switches 128
6.4.3 Separating Matching into Input and Output Arbitrations 130
6.4.4 Weighted Arbitration Schemes 130
6.4.5 Arbitration Schemes based on Round-robin Selection 135
6.5 CICB Switches with Internal Variable-length Packets 141
6.6 Output Emulation by CICB Switches 141
6.7 Conclusions 144

7 Dual Scheduling Algorithm in a Generalized Switch: Asymptotic Optimality and Throughput Optimality 147
Lijun Chen, Steven H. Low, and John C. Doyle
7.1 Introduction 148
7.2 System Model 150
7.2.1 Queue Length Dynamics 151
7.2.2 Dual Scheduling Algorithm 152
7.3 Asymptotic Optimality and Fairness 153
7.3.1 An Ideal Reference System 153
7.3.2 Stochastic Stability 154
7.3.3 Asymptotic Optimality and Fairness 155
7.4 Throughput-optimal Scheduling 159
7.4.1 Throughput Optimality and Fairness 159
7.4.2 Optimality Proof 160
7.4.3 Flows with Exponentially Distributed Size 163
7.5 A New Scheduling Architecture 165
7.6 Conclusions 166

8 The Combined Input and Crosspoint Queued Switch 169
Kenji Yoshigoe and Ken Christensen
8.1 Introduction 169
8.2 History of the CICQ Switch 172
8.3 Performance of CICQ Cell Switching 175
8.3.1 Traffic Models 176

8.3.2 Simulation Experiments ... 177
8.4 Performance of CICQ Packet Switching ... 179
8.4.1 Traffic Models ... 179
8.4.2 Simulation Experiments ... 179
8.5 Design of Fast Round-robin Arbiters ... 181
8.5.1 Existing RR Arbiter Designs ... 182
8.5.2 A New Short-term Fair RR Arbiter – The Masked Priority Encoder (MPE) ... 183
8.5.3 A New Fast Long-term Fair RR Arbiter – The Overlapped RR (ORR) Arbiter ... 186
8.6 Future Directions – The CICQ with VCQ ... 188
8.6.1 Design of Virtual Crosspoint Queueing (VCQ) ... 189
8.6.2 Evaluation of CICQ Cell Switch with VCQ ... 190
8.7 Summary ... 192

9 Time–Space Label Switching Protocol (TSL-SP) ... 197
Anpeng Huang, Biswanath Mukherjee, Linzhen Xie, and Zhengbin Li
9.1 Introduction ... 197
9.2 Time Label ... 198
9.3 Space Label ... 200
9.4 Time–Space Label Switching Protocol (TSL-SP) ... 201
9.5 Illustrative Results ... 205
9.6 Summary ... 209

10 Hybrid Open Hash Tables for Network Processors ... 211
Dale Parson, Qing Ye, and Liang Cheng
10.1 Introduction ... 211
10.2 Conventional Hash Algorithms ... 213
10.2.1 Chained Hash Tables ... 214
10.2.2 Open Hash Tables ... 215
10.3 Performance Degradation Problem ... 216
10.3.1 Improvements ... 218
10.4 Hybrid Open Hash Tables ... 219
10.4.1 Basic Operations ... 219
10.4.2 Basic Ideas ... 219
10.4.3 Performance Evaluation ... 220
10.5 Hybrid Open Hash Table Enhancement ... 222
10.5.1 Flaws of Hybrid Open Hash Table ... 222
10.5.2 Dynamic Enhancement ... 223
10.5.3 Adaptative Enhancement ... 224
10.5.4 Timeout Enhancement ... 224
10.5.5 Performance Evaluation ... 224
10.6 Extended Discussions of Concurrency Issues ... 225
10.6.1 Insertion ... 225
10.6.2 Clean-to-copy Phase Change ... 226
10.6.2 Timestamps ... 227
10.7 Conclusion ... 227

Index ... 229

List of Contributors

Nirwan Ansari
Department of Electrical & Computer Engineering
New Jersey Institute of Technology
e-mail: nirwan.ansari@njit.edu

Nicholas Bambos
Electrical Engineering and Management Science & Engineering Departments
Stanford University
e-mail: bambos@stanford.edu

Andrea Bianco
Dipartimento di Elettronica
Politecnico di Torino
C.so Duca degli Abruzzi 24
Torino, Italy
e-mail: Andrea.bianco@polito.it

Lijun Chen
Engineering and Applied Science Division
California Institute of Technology
Pasadena, CA 91125, USA
e-mail: chen@cds.caltech.edu

Liang Cheng
Laboratory of Networking Group
Computer Science and Engineering Department
Lehigh University
Bethlehem, PA 18015
e-mail: cheng@cse.lehigh.edu

Ken Christensen
Department of Computer Science and Engineering
University of South Florida
Tampa, FL 33620
e-mail: christen@cse.usf.edu

John C. Doyle
Engineering and Applied Science Division
California Institute of Technology
Pasadena, CA 91125, USA
e-mail: doyle@cds.caltech.edu

Itamar Elhanany
Electrical & Computer Engineering Department
The University of Tennessee
Knoxville, TN 37996-2100
e-mail: itamar@ieee.org

Paolo Giaccone
Dipartimento di Elettronica
Politecnico di Torino
C.so Duca degli Abruzzi 24
Torino, Italy
e-mail: Paolo.Giaccone@polito.it

Zhen Guo
Department of Electrical & Computer Engineering
New Jersey Institute of Technology
e-mail: zhen.guo@njit.edu

Mounir Hamdi
Department of Computer Science
The Hong-Kong University of Science and Technology
e-mail: hamdi@cs.ust.hk

Anpeng Huang
Department of Computer Science
The University of California at Davis
e-mail: hapku@cs.ucdavis.edu

Emilio Leonardi
Dipartimento di Elettronica
Politecnico di Torino
C.so Duca degli Abruzzi 24
Torino, Italy
e-mail: Emilio.Leonardi@polito.it

Zhengbin Li
Department of Computer Science
The University of California at Davis
e-mail: lizhb@cs.ucdavis.edu

Xin Li
Department of Computer Science
The Hong-Kong University of Science and Technology
e-mail: lixin@cs.ust.hk

Jing Liu
Department of Computer Science
The Hong-Kong University of Science and Technology
e-mail: liujing@cs.ust.hk

Steven H. Low
Engineering and Applied Science Division
California Institute of Technology
Pasadena, CA 91125, USA
e-mail: slow@cds.caltech.edu

Brad Matthews
Electrical & Computer Engineering Department
The University of Tennessee
Knoxville, TN 37996-2100
e-mail: bradmatthews@ieee.org

Marco Mellia
Dipartimento di Elettronica
Politecnico di Torino
C.so Duca degli Abruzzi 24
Torino, Italy
e-mail: Marco Mellia@polito.it

Lotfi Mhamdi
Department of Computer Science
The Hong-Kong University of Science and Technology
e-mail: lotfi@cs.ust.hk

Biswanath Mukherjee
Department of Computer Science
University of California at Davis
e-mail: mukherje@cs.ucdavis.edu

Fabio Neri
Dipartimento di Elettronica
Politecnico di Torino
C.so Duca degli Abruzzi 24
Torino, Italy
e-mail: Fabio.Neri@polito.it

Dale Parson
Agere Systems
Allentown , PA 18019
e-mail: dparson@agere.com

Konghong Pun
Department of Computer Science
The Hong-Kong University of Science and Technology
e-mail: konghong@cs.ust.hk

Roberto Rojas-Cessa
Department of Electrical & Computer Engineering
New Jersey Institute of Technology
e-mail: rojasces@njit.edu

Kevin Ross
University of California Santa Cruz
Technology and Information Management
e-mail: kross@soe.ucsc.edu

Vahid Tabatabaee
Institute for Advanced Computer Studies
University of Maryland at College Park
e-mail: vahid@eng.umd.edu

Linzhen Xie
Department of Computer Science
University of California at Davis
e-mail: tydxlz@cs.ucdavis.edu

Qing Ye
Laboratory of Networking Group
Computer Science and Engineering Department
Lehigh University
Bethlehem, PA 18015
e-mail: qiy3@lehigh.edu

Kenji Yoshigoe
Department of Computer Science
University of Arkansas at Little Rock
Little Rock, AR 72204
e-mail: kxyoshigoe@ualr.edu

1

Architectures of Internet Switches and Routers

Xin Li, Lotfi Mhamdi, Jing Liu, Konghong Pun, and Mounir Hamdi

The Hong-Kong University of Science & Technology. {lixin,lotfi,liujing,
konghong,hamdi}@cs.ust.hk

1.1 Introduction

Over the years, different architectures have been investigated for the design and implementation of high-performance switches. Particular architectures were determined by a number of factors based on performance, flexibility and available technology. Design differences were mainly a variation in the queuing functions and the switch core. The crossbar-based architecture is perhaps the dominant architecture for today's high-performance packet switches (IP routers, ATM switches, and Ethernet switches) and owes its popularity to its scalability (when compared to the shared-bus/shared-memory architectures), efficient operation (supports multiple I/O transactions simultaneously) and simple hardware requirements. The architecture includes the input-queued (IQ) crossbar fabric switch with its variations (Output-queued, OQ, switch and Combined Input–Output-queued, CIOQ, switch) and the internally buffered crossbar fabric switch (BCS).

IQ switches have gained much interest in both academia and industry because of their low cost and scalability. The IQ switch has a low internal speedup because the crossbar fabric has the same speed as that of the external line. Although the head-of-line (HoL) blocking problem limits the achievable throughput of an IQ switch to approximately 58.6% [1], the well-known virtual output queuing (VOQ) architecture [2] was proposed and has improved switching performance by several orders of magnitude, making IQ switches more desirable. However, the adoption of VOQ has created a more serious problem, namely, the centralized scheduler. An arbitration algorithm examines the contents of all the input queues, and finds a conflict-free match between inputs and outputs. The well-known optimal algorithms (*i.e.* maximum-weight-matching or MWM) are too complex to implement at high speed while the iterative algorithms, proposed as an alternative to the MWM algorithms, fail to perform well under real world input traffic conditions.

As bufferless scheduling algorithms reach their practical limitations due to higher port numbers and data rates, internally buffered crossbar switches (BCS) have started to attract researchers because of the great potential they have in solving the complexity and scalability issues faced by their bufferless predecessors. The increasing demand for terabit switches and routers means that future commercial packet switches must be implemented with reduced scheduling complexity. The buffered crossbar architecture can inherently implement distributed scheduling schemes and has been considered a viable alternative to bufferless crossbar switches to improve performance. The presence of internal buffers drastically improves the overall performance of the switch as it offers two distinct advantages. First, the adoption of internal buffers makes the scheduling totally distributed, dramatically reducing the arbitration complexity. Second, and most importantly, these internal buffers reduce (or avoid) output contention as they allow the inputs to make cell transfers concurrently to a single output. While there have been many architectures for the (BCS) [3, 4, 5], our focus in this chapter is on BCS with VOQs (denoted by VOQ/BCS).

Despite the advantages that the VOQ/BCS architecture offers with regards to the bufferless architecture, both are seen as unscalable and unable to keep up with Internet growth in the foreseeable future. In fact, with progress in wavelength division multiplexing (WDM) technology and optical fiber transmission, switches and routers are becoming the bottleneck of the overall system. The increasing data rates on the Internet are causing a scalability challenge for these crossbar-based architectures. This scalability limitation can be attributed to the nature of crossbar-based fabric, as a quadratic growth in the number of crosspoints puts a limit on chip area and/or pin counts. As a result, there is an urgent need for truly scalable high-performance packet switches that has motivated researchers and industry to look for alternative switch fabrics. In particular, architectures such as the Clos-network switch architecture and the optical-electronic switch architecture are starting to receive attention because of their great potential in entirely solving the scalability issues.

The Clos-network switch architecture is the most popular example among numerous examples of multi-stage switches in the communication world. It is mainly a three-stage switch as shown in Figure 1.11. There are switching elements (SE) at each stage which are connected. While the SEs can be any interconnect, typically they are crossbars of smaller sizes. The SE at each stage of the Clos network is capable of transmitting a packet from an input port to any of the output ports. Clos-network switches can be classified as buffered or bufferless, where the former uses buffers to store packets in the second-stage SEs and the latter does not [6, 7]. Buffers in the second-stage SEs can help resolve contention among packets from different first-stage modules but may cause a sequencing problem. While the Clos-network switches have been proposed for a some time now, relatively little research (especially compared to crossbars) has been conducted on making them scalable and high-speed packet switches that can serve the needs of the Internet. One of the goals of the current survey is to fill this gap.

Given the effort put by both industry and academia into the all-electronic switch/router design, numerous studies show that all-electronic technology can no longer be a viable solution for the design of scalable switches/routers (*i.e.* 256×256 and beyond). Alongside the scalability challenge, an electronic switch fabric becomes very costly beyond a reasonable size (*e.g.* 128×128) and data rate (*e.g.* 10 Gb/s) due to the ever-increasing power and chip count requirements for its implementation. Moreover, current WDM systems offer 32–64 wavelengths at 2.5–10Gb/s/wavelength, approaching a 1 Tb/s capacity, while research-level systems already exceed multiterabits in a single fiber. As a result, traffic growth will not be limited by fiber bandwidth and/or optical components in the links. The total data rate of a single fiber is increasing at a rate faster than the switching and routing equipment that terminates and switches traffic at a carrier's central office or point of presence (POP). In particular, switches and routers are becoming the bottlenecks of the overall system. While there has been a lot of attention paid to all-optical switches as a solution to this bottleneck problem, most current POP switching equipment is electronic switches/routers with optical I/O, and an all-optical solution in not expected to be viable in the foreseeable future. For this reason, research focus is shifting towards a solution that takes advantage of the strengths of both electronics and optics, with the ultimate goal of designing practical switches and routers that can scale with the Internet traffic growth as well as keep up with the advances in WDM fiber transmission. Recently, attention has been given to hybrid optical–electronic architectures where the linecards and switch scheduler are designed using electronics and the switch fabric is designed with optical technology. This switch architecture provides advantages such as scalability, lower power consumption, and lower cost. However, a hybrid opto-electronic packet switch/router presents a unique challenge: the reconfiguration time of an optical switch fabric is much longer than that of an electronic fabric, and the end-to-end clock recovery in such a system adds to the reconfiguration overhead.

The goal of this chapter is to survey each of the above presented architectures, and will be organized as follows. Section 1.2 presents the bufferless switching architecture with its different types and scheduling schemes. Section 1.3 presents the buffered crossbar switch architecture and discusses its scheduling processes and proposed schemes. In Section 1.4, we present the Clos-network architecture and discuss its variants and scheduling. Section 1.5 presents the optical switching trend and the potential for building such cost-effective and highly scalable switches. Section 1.6 concludes the paper and suggests problems for further research.

1.2 Bufferless Crossbar Switches

1.2.1 Introduction to Switch Fabrics

Crossbar switch fabric is an active non-blocking switch fabric. In crossbar architecture, each line card is connected by a dedicated point-to-point link to the central switch fabric. The inputs and outputs are connected at switching points, called crosspoints. It is the scheduler's responsibility to set configurations for the crosspoint matrix and for each configuration. One input/output can only be connected to one output/input. The basic architecture of a crossbar switch is shown in Figure 1.1.

For bufferless crossbar switches, there is no buffer at the crosspoint. However, buffers may be placed at the input side, output side, or both. Based on where the buffers are placed, bufferless crossbar switches are categorized into input-queued switches, output-queued switches, and combined input–output queued switches.

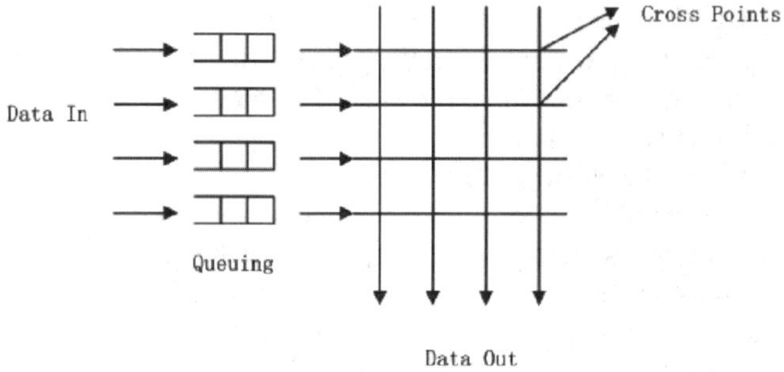

Figure 1.1. Crossbar architecture

1.2.2 Output-queued Switches

Traditionally, switches and routers have been most often designed with an output queuing strategy. This strategy has advantages in that guaranteed qualities-of-service (QoS) can be provided for the system, which is able to control packet departure times [8, 9]. An output-queued switch is attractive as it can always achieve 100% throughput. However, since there are no queues at the inputs, all arriving cells must be immediately delivered to their outputs. The necessary simultaneous delivery of all arriving cells to the outputs becomes a disadvantage if the requirements for internal interconnection bandwidth and memory bandwidth are too great. For a switch with N input ports, there can be up to N cells, one from each input, arriving at any one output simultaneously. Thus, in order to receive all the cells at one time the memory needs a bandwidth of N cell time write accesses. This requirement is referred to as the internal speedup of the switch [10], so an output-queued switch has an internal speedup of N. The current demand for bandwidth is growing rapidly and as switch sizes continue to increase, memory bandwidth will be insufficient for output queuing to be practical.

1.2.3 Input-queued Switches

Memory bandwidth is not a problem with the input queuing strategy. In input-queued switches, arriving cells are buffered on the input side and extracted to pass through the switch fabric according to some arbitration algorithm. Contention within the switch fabric and input/output interfaces is resolved by the arbitration algorithm (each input can deliver at most one cell to the switch in one cell time and each output can accept no more than one cell in one cell time).

Queuing Strategies

- *Single FIFO and Head-of-line Blocking*: It is relatively easy to implement single FIFO (first-in-first-out) switches. A single FIFO queue is used at each input where only the first cell in each queue is eligible to be forwarded. Head-of-line (HoL) blocking may occur where no cell can be transmitted from an input because of a contention of the first packet in the FIFO queue. Though the next cell in the queue may be without contention, it cannot be transmitted as this would disrupt the queue. Single FIFO switches suffering from HoL blocking result in poor performance. It is well known that such a switch with Bernoulli I.I.D. arrivals under uniform traffic can only achieve a maximum throughput of 58.6% when the number of ports is large [1]. For periodic traffic, HoL blocking can lead to even worse performance [11].

- *Windowing Mechanism*: This mechanism enables the scheduler of the switch to look ahead w (window size) cells in the queue at a time, relieving the HoL blocking problem and increasing performance by allowing cells behind the head of the queue to be extracted. If the cell at the head of the queue cannot be transferred to the intended output because of contention, the second cell is considered, and so on, up to the wth queue position. Note that when $w = 1$, it is single FIFO queuing. The performance of input-queued switches adopting this queue organization grows with w, however, this gain is limited as the complexity of the queuing mechanism increases.
- *Virtual Output Queuing*: HoL blocking can be completely eliminated by using a virtual output queuing [2] architecture at the input side. Rather than maintaining a single FIFO queue for all cells, each input maintains a separate queue for each output as shown in Figure 1.2. There are thus a total of N^2 input queues, where each separate queue is called a VOQ and operates according to the FIFO discipline. The scheduler will select among the HoL cells of each VOQ and transmit them. HoL blocking is eliminated as no cell can be held up by a cell ahead of it that is destined for a different output. When virtual output queuing is employed, the performance of the switch depends on the scheduling algorithm that decides which cells should be transmitted during a cell time under the condition that only one cell can be delivered to each input and only one cell can be accepted at each output. With suitable scheduling algorithms, an input-queued switch using virtual output queuing can increase the throughput from 58.6% to 100% for both uniform and non-uniform traffic [12].

1.2.4 Scheduling Algorithms for VOQ Switches

Per one time slot, each input can send at most one cell and each output can receive at most one cell. Scheduling algorithms are necessary for crossbar switches to find a proper one-to-one match between inputs and outputs in order to configure the crossbar. A variety of scheduling algorithms are proposed for the VOQ architecture. This section presents an overview of some popular and effective schemes.

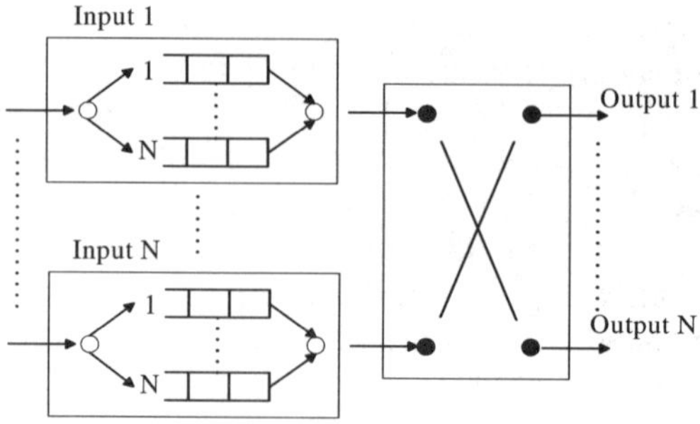

Figure 1.2. Virtual output queuing

Maximum Weight Matching Scheduling Algorithms

In maximum weight matching algorithms, a weight is attached with each request from inputs to outputs. The weight could be the number of cells in the VOQ, the waiting time of the HoL cell in the VOQ, etc.

- *LQF*: The longest queue first (LQF) algorithm [13] takes the number of cells in each VOQ as weight. The algorithm picks a match such that the sum of served queues' lengths is maximized. LQF can lead to the starvation of some queues. Because it does not consider the waiting time of cells, queues with short length may be starved even though the wait time of their HoL cells surpasses the total weight time experienced by cells in a longer queue.
- *OCF*: The oldest cell first (OCF) algorithm [13] uses the waiting times of HoL cells as weights. The OCF algorithm selects a match such that the sum of all served queues' waiting time is maximized. Unlike the LQF algorithm, the OCF algorithm does not starve any queue and unserved HoL cells will eventually become old enough to be served.
- *LPF*: Both LQF and OCF have high computation complexity of $O(N^3 \log N)$ and it is impractical to implement them in hardware. The longest port first (LPF) algorithm [14] is designed to overcome the complexity problem of LQF and OCF and can be implemented in hardware at high speed. The weight of the LPF algorithm is the total number of cells queued at the input and output interfaces: $w_{ij} = \sum_{k=1}^{N} L_{i,k} + \sum_{k=1}^{N} L_{k,j}$ (L_{ij} is the number of cells in VOQ_{ij}). This sum is called port occupancy, which represents the workload or congestion that a cell faces as it competes for transmission. It is proved in [14] that the maximum weight match found by LPF is also a maximum size match. Thus, a modified maximum size matching algorithm, which makes LPF less complex than LQF, is used to implement LPF.

MWM scheduling algorithms achieve 100% throughput under any admissible traffic. However, their good performance and stability come at the expense of high computation complexity.

Approximating Maximum Size Matching Scheduling Algorithms

Approximating maximum size matching algorithms are fast and simple to implement in hardware with today's technologies. They provide 100% throughput under uniform traffic and fairly good delay performance as well. However, they are not stable under non-uniform traffic. Most of the approximating maximum size matching algorithms are iterative algorithms.

- *PIM*: The Parallel Iterative Matching (PIM) [15] algorithm attempts to approximate a maximum size matching algorithm by iteratively matching the inputs to the outputs until it finds a maximum size match. Each iteration consists of three steps:
  Step 1. *Request.* Each unmatched input sends a request to every output for which it has a queued cell.
  Step 2. *Grant.* If an unmatched output receives any requests, it grants one by randomly selecting from all requests.
  Step 3. *Accept.* If an input receives more than one grant, it accepts one by random selection.
  The PIM algorithm faces some problems. First, randomness is difficult and expensive to implement at high speed. Each arbiter must make a random selection among the members of a time-varying set. Second, when the switch is oversubscribed, PIM can lead to unfairness between connections. Finally, PIM does not perform well for a single iteration: it limits the throughput to approximately 63%, only slightly higher than a FIFO switch.

- *Round-robin scheduling*: All existing round-robin scheduling algorithms have the same three steps as PIM. Instead of randomly matching cells, the input and output arbiter adopts a round-robin scheme where the inputs and outputs take turns. Round-robin scheduling overcomes two problems in PIM: complexity and unfairness. Implemented as priority encoders, the round-robin arbiters are much simpler and can perform faster than random arbiters. The rotating scheme makes the algorithm assign bandwidth equally and more fairly among requests.

  - iSLIP: One iteration of iSLIP [2] consists of three steps:
    Step 1. *Request.* Each unmatched input sends a request to every output for which it has a queued cell.
    Step 2. *Grant.* If an output receives any requests, it chooses the one that appears next in a fixed, round-robin schedule starting from the highest priority element. The output notifies each input whether or not its request was granted. The pointer to the highest priority element of the round-robin schedule is incremented (modulo N) to one location beyond the granted input if and only if the grant is accepted in Step 3. If no request is received, the pointer stays unchanged.
    Step 3. *Accept.* If an input receives a grant, it accepts the one that appears next in a fixed round-robin schedule starting from the highest priority element. The pointer to the highest priority element of the round-robin schedule

is incremented (modulo N) to one location beyond the accepted one. If no grant is received, the pointer stays unchanged. iSLIP updates the grant pointers only when the grant is accepted. In this scheme, starvation is avoided.

- FIRM: FCFS in round-robin matching (FIRM) [16] achieves improved fairness (as it approximates FCFS) and has tighter service guarantee than iSLIP. The only difference between iSLIP and FIRM lies in Step 2. FIRM moves the grant pointer regardless of whether the grant is accepted or not, *i.e.*:
Step 2. *Grant.* If an output receives any requests, it chooses the one that appears next in a fixed, round-robin schedule starting from the highest priority element. The output notifies each input whether or not its request was granted. The pointer to the highest priority element of the round-robin schedule is incremented (modulo N) to one location beyond the granted input if and only if the grant is accepted in Step 3. If the grant is not accepted, the pointer is placed to the granted input. If no request is received, the pointer stays unchanged.

The modification of the grant pointer results in enhanced fairness in terms of FCFS service, *i.e.* FIRM approximates FCFS closer than iSLIP with the use of the round-robin pointers. FIRM achieves this as it forces the scheduler to issue the next grant to the unsuccessful request until the grant is accepted while iSLIP may take the newly arrived request first.

Randomized Scheduling Algorithms

The motivation for proposing randomized algorithms is to overcome the complexity of MWM algorithms and to achieve stability under any admissible traffic.

- TASS: TASS [17] is the basic randomized algorithm proposed by Tassiulus. The steps for this algorithm are as follows:
(a) Let $S(t)$ be the schedule used at time t.
(b) At time $t+1$ choose a match $R(t+1)$ uniformly at random from the set of all $N!$ possible matches.
(c) Let $S(t+1) = arg\max < S, Q(t+1) >$ ($Q(t+1)$ is the queue-lengths matrix at time $t+1$.)
It was proven that if the probability of $R(t+1)$ being equivalent to the maximum weight matching is lower bounded by some constant c, the algorithm is stable. Although TASS achieves stability under any admissible traffic, it does not have good delay performance, especially under non-uniform traffic. Recently, a group of randomized algorithms including APSARA, LAURA, and SERENA were proposed [18]. The motivation behind these algorithms is to improve the delay performance by exploiting some features of the switching problem while maintaining stability.

- APSARA: APSARA [18] selects a match for the current time slot based on three inputs: the match of the last time slot $S(t)$, neighbors of the match $S(t)$, and a randomly chosen match $Z(t+1)$. A neighbor of a match S, denoted as $N(S(t))$, is obtained by exchanging the input/output pair of two connections in the match. The weight is then computed from these inputs and chosen as the maximum match time at $t+1$. The neighbor is used here with the objective of searching the space matches in parallel.

- LAURA: LAURA [18] differs from APSARA in that LAURA uses a non-uniform random sampling which has the purpose of keeping heavy edges in the matching procedure and a corresponding merging procedure for weight augmentation. By merging the matches of the randomly chosen $Z(t+1)$ with the match of last time slot $S(t)$, a match with increased weight is obtained. Notice that when compared with TASS, 'merge' is used instead of 'max' due to the performance improvement obtained with the merging algorithms.

- SERENA: SERENA [18] makes some modifications to LAURA where it uses the arrival information in the merging procedure. The arrival graph, $A(t+1)$, is one component in the merging procedure and is directly used when it is a match. Otherwise, modification to $A(t+1)$ is required in order to yield a match. This information is used based on the observation that the accumulation of queue lengths is due to arrival events, which also serve as a source of randomness.

Performance Comparison among Algorithms

MWM algorithms perform very well in terms of both delay and stability. However, the computation complexity required to implement them is generally too high to be practical. The more practical approximating MSM algorithms perform well under uniform traffic, but are not stable under non-uniform traffic. Randomized algorithms are linear in complexity and provide the benefit of being stable under any admissible traffic as well. However, the delay encountered is higher than that of approximating MSM algorithms, as randomized algorithms have been designed with objectives of stability rather than small average delay. Figure 1.3 shows the average delay under uniform traffic for these typical algorithms. Figure 1.4 compares the average delay of the same algorithms under diagonal traffic.

1.2.5 Combined Input–Output-queued Switches

VOQs and an increase in the internal speedup of a switch are used to solve the HoL blocking problem. If the switch fabric has an internal speedup, *i.e.* a few times faster than the line rate, buffers are required at both input and output sides. This is a combined input–output queued (CIOQ) switch, which can emulate an OQ switch with a small speedup. There are two emulation schemes: emulating an OQ switch on throughput and emulating both throughput and cell delivery order.

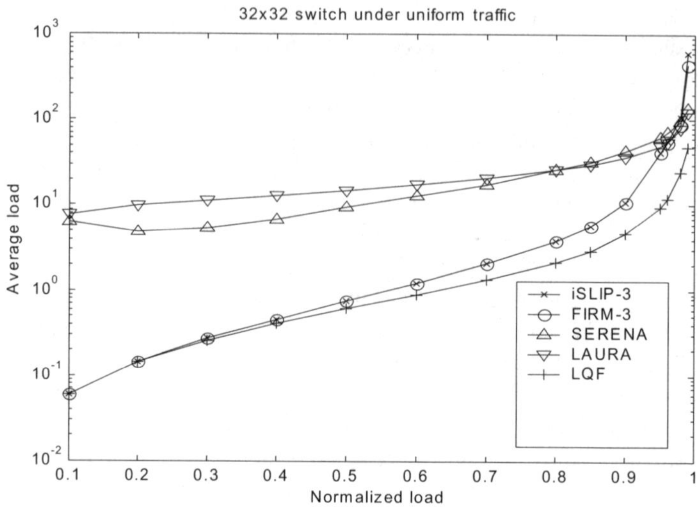

Figure 1.3. Average delay performance for iSLIP (run with three iterations), FIRM (run with three iterations, SERENA, LAURA and LQF under uniform traffic

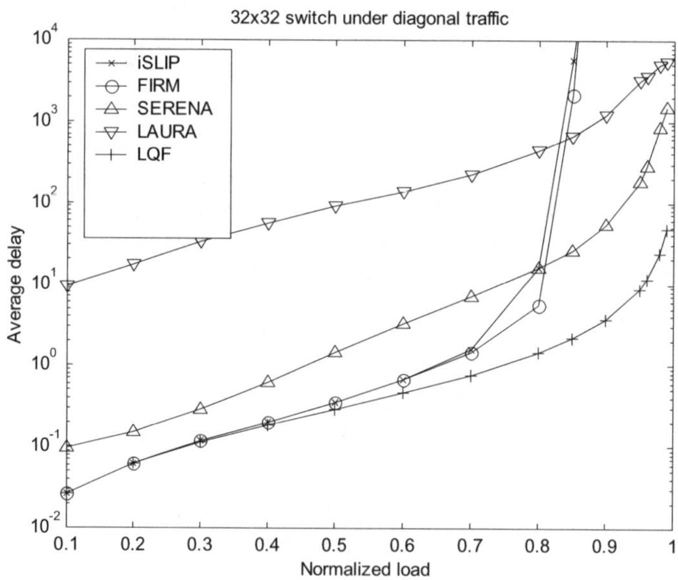

Figure 1.4. Average delay performance for iSLIP, FIRM, SERENA, LAURA and LQF under diagonal traffic

- Exact Emulation of an OQ Switch

Consider an OQ switch whose output buffers are FIFO. A CIOQ switch is said to behave identically to an OQ switch if, under identical inputs, the departure time of every cell from both switches is identical. This is also called the exact emulation of the OQ switch.

It was first formulated by Prabhakar and Mckeown in [10] that a CIOQ switch with a speedup of 4 can behave identically to a FIFO-OQ switch. This result holds for switches with an arbitrary number of ports and for any traffic arrival pattern. Their proof is based on the scheduling algorithm called Most Urgent Cell First (MUCFA). The urgency value of a cell used in the MUCFA is defined as the distance it clones from the head of the output buffer in the shadow switch. The cells in any output buffer of the CIOQ switch are arranged in increasing order of urgencies, with the most urgent cell at the head. Once a cell is forwarded to its output in the CIOQ switch, its position is determined by its urgency. MUCFA works as follows:

(1) At the beginning of each phase, each output obtains its most urgent cells from the inputs.

(2) If more than one output requests an input, then the input will grant the output whose cell has the smallest urgency number. Ties are broken by the smallest port number.

(3) Each unmatched output obtains its next most urgent cell from another unmatched input. Then go to step 2.

(4) When the matching of inputs and outputs is no longer possible, cells are transferred and MUCFA goes to step 1 for the next phase.

The way in which MUCFA matches inputs and outputs is a variation of the stable marriage problem, which was first introduced by Gale and Shapley and can be solved by a well-known algorithm called GSA. GSA can find a stable match in N^2 iterations [19]. MUCFA is extended to achieve the exact emulation by a speedup of only 2 by a new algorithm called Joined Preferred Matching (JPM) algorithm [20]. Similarly, Critical Cell First (CCF) has been proposed to emulate an OQ switch with speedup 2 and with different output scheduling disciplines [10]. There are two main disadvantages for those GSA-based algorithms mentioned above. First, the stable match in each phase can take as many as N^2 iterations. Second, the algorithms have a high information complexity: they need to maintain a large amount information which is not locally available. The DTC strategy is proposed to reduce the number of iterations and the GBVOQ algorithm is proposed to avoid using global information [10]. However, these two solutions cannot be combined.

- Work-conserving

A switch is work-conserving if and only if each output is active at the end of the time slot T, where there are cells (either at input or at the output buffers) at the beginning of that time slot. A work-conserving switch provides the same

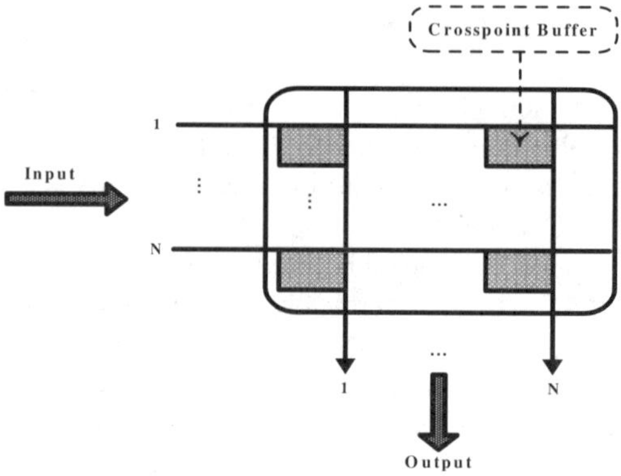

Figure 1.5. Pure buffered crossbar architecture

throughput performance as an OQ switch. The Lowest Occupancy Output First Algorithm (LOOFA) has been proposed in [21] with the work-conserving property in a CIOQ switch with a speedup of 2. An integer variable, occupancy, is associated with each output queue in LOOFA. The occupancy of an input j at any time is simply the number of cells currently residing in output j's buffer.
(1) Each unmatched input selects the non-empty VOQ going to an unmatched output with the lowest occupancy and sends a request to that output.
(2) The output, upon receiving requests from multiple inputs, selects one and sends a grant to that input.
(3) The switch returns to step 1 until no more matches can be made.
This algorithm essentially gives priority to output channels with low occupancy, thereby attempting to simultaneously maintain work conservation across all output channels. The work conserving feature of the switch is independent of the selection algorithm used at the outputs.

1.3 Buffered Crossbar Switches

1.3.1 Buffered Crossbar Switches Overview

For many years, buffered crossbar switches have been considered a viable solution to improve the switching throughput as an alternative to bufferless crossbar switches. Buffered crossbar switches have been studied for at least two decades [4, 22, 23]. In an architecture called *pure* buffered crossbar, as shown in Figure 1.5, buffering occurs *exclusively* at the crosspoints and is utilized to minimize cell loss. The number of ports is limited by the memory amount that can be implemented in a module chip. An example of this architecture was proposed in [4], where a 2×2 switch module with a crosspoint memory of 16 Kbytes each was implemented.

Unfortunately, at that time, it was not possible to embed enough buffering on-chip and therefore this architecture was unable to comply with the required cell loss rate. In order to overcome the excessive on-chip memory requirement, buffered crossbar switches with input queues were proposed [23, 24]. While there have been many studies analyzing the performance of input FIFO based buffered crossbar switches, a recent result proved that a buffered crossbar employing FIFO queues as its input queuing can achieve 100% throughput under uniform traffic arrivals [25]. As with traditional IQ switches, buffered crossbar switches with input VOQs were researched in order to improve the switching performance.

Buffered crossbar switches with input VOQs (denoted VOQ/BCS) have recently received a lot of interest from both research and industrial communities. Numerous research groups have studied the VOQ/BCS architecture [3, 5, 26]. A representative example from industry that implemented VOQ/BCS is [27]. The first fixed-size-cell VOQ/BCS architecture was introduced in [3] and was shown to outperform conventional IQ switching performance. A Combined Input-One-cell-Crosspoint Buffered Switch (CIXB-1) with VOQs at the input employing a round-robin arbitration scheme was proposed in [28], and was shown to achieve 100% throughput under uniform traffic. The same architecture was extended in [29] to support more than a one cell internally buffered crossbar with input VOQs along with round-robin arbitration. This architecture shows improvement over that of [28] by achieving 100% under uniform traffic as well as non-uniform traffic. VOQ/BCS architecture operating on variable length packets has been studied and was introduced in [30, 31]. When directly operating on variable length packets, a VOQ/BCS no longer requires a segmentation and reassembly (SAR) circuitry and no speedup to compensate for it [31].

VOQ/BCS have also been studied in an attempt to emulate OQ switching. Very interesting results have been proposed showing that a VOQ/IBC, running twice as fast as the line rate, can emulate a FIFO OQ switch [32]. Other results [33], proved that a VOQ/IBC with a speedup of two can emulate a broad class of algorithms (*i.e.*, FCFS, Strict Priority, Early Deadline First) operating on an OQ switch. A more recent and extended result [26] showed that a VOQ/IBC switch with two times speedup can provide 100% throughput, rate and delay guarantees.

As will be shown, the VOQ/BCS has many advantages that alleviate the scheduling task and make it simple. The increasing demand for terabit switches and routers means that future commercial packet switches must be implemented with reduced scheduling complexity, and buffered crossbar architecture, inherently, can implement distributed scheduling schemes. In fact, buffered crossbar switches have been considered as a viable alternative to bufferless crossbar switches to improve the switching performance.

1.3.2 The VOQ/BCS Architecture

In this section, we present the buffered crossbar architecture (VOQ/BCS) using the notation and terminology that will be referred to in the rest of this chapter. We present the switch model and its components such as inputs, internally buffered crossbar fabric, and outputs. Then, we present the three steps that a scheduling cycle consists of: input scheduling, output scheduling, and delivery notification.

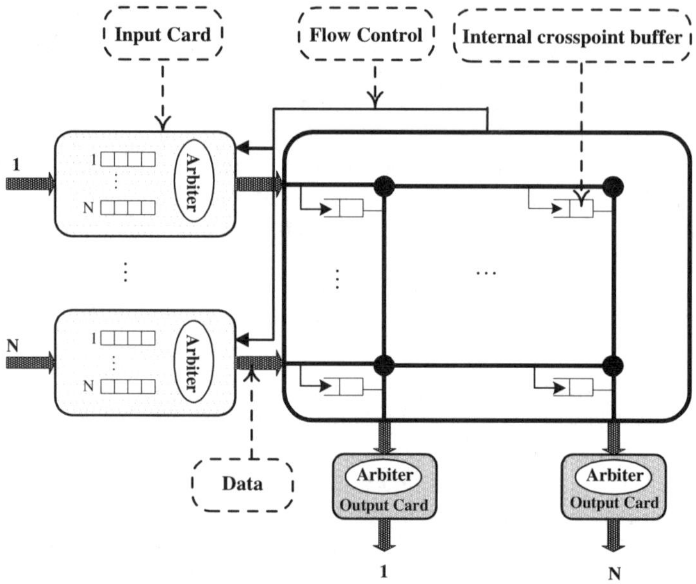

Figure 1.6. The VOQ/BCS architecture

Architecture and Terminology

As shown in Figure 1.6, an $N \times N$ VOQ/BCS switch architecture consists of N input cards, with each maintaining N VOQs, one for each output. The buffered crossbar fabric component is the main characteristic of thc VOQ/BCS that differentiates it from the conventional IQ/VOQ architecture, where small buffers are added per crosspoint. Fixed size packets, or cells, are considered. Upon their arrival at the switch, variable length packets are segmented into cells for internal processing and re-assembled before they leave the switch. A processing cycle has a fixed length, called cell or time slot. The architecture can be divided into three major parts and described as follows:

- **Input Cards:** There are N input cards; each one maintains N logically separated VOQs. When a packet destined to output j, $0 \leq j \leq N-1$, arrives at the input card i, $0 \leq i \leq N-1$, it is held in VOQ_{ij}. A VOQ_{ij} is said to be eligible for being scheduled in the input scheduling process if it is not empty and the internal buffer XP_{ij} is empty.
- **Buffered Fabric:** The internal fabric consists of N^2 buffered crosspoints (XP). Each crosspoint has a one-cell buffer. A crosspoint XP_{ij}, holds cells coming from input i and going to output j.
  XPB_i is the set of the internal buffers for all outputs j ($XP_{i0}+\cdots+XP_{i(N-1)}$).
  XPB_j is the set of the internal buffers for all inputs i ($XP_{0j}+\cdots+XP_{(N-1)j}$).
- **Flow Control:** Between each two time slots a flow control mechanism is performed for information exchange between the crosspoints and the input cards.

(a) Sequential

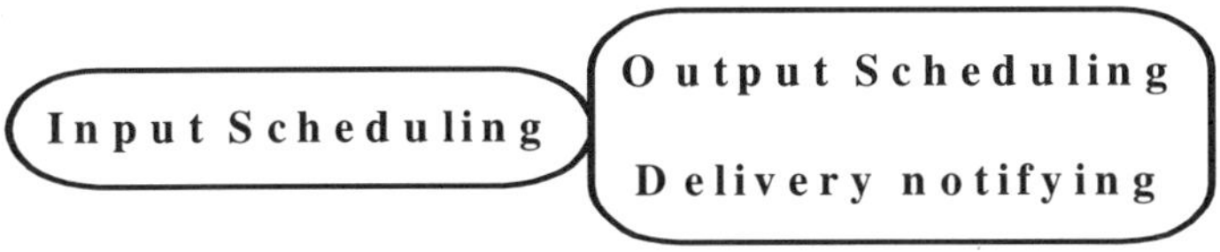

(b) Overlapped

Figure 1.7. Scheduling cycle for the VOQ/BCS architecture

Each crosspoint XP_{ij} tells its corresponding input card i whether or not it is ready to receive a cell during the next time slot.

It is of note that the arbiters at the inputs and the outputs are totally distributed. There are as many arbiters as input ports, and the same applies for the outputs. The arbitration made by any arbiter does not depend on the other arbiter's decisions. At every input port, all an arbiter i needs to maintain is the state of the VOQs belonging to its port i and the internal buffer's XPB_i state corresponding to these VOQs. However, for each output arbiter, it is even simpler. All an output arbiter j needs to maintain is the occupancy of its corresponding internal buffers XPB_j.

Scheduling Process

With the structure described above and the corresponding notation, a scheduling cycle consists of three steps:

1. **Input scheduling:** each input selects one cell, in a predefined manner, from the HoL of an eligible VOQ.
2. **Output scheduling:** each output selects one cell, in a predefined manner, from all the internally buffered cells in the crossbar to be delivered to the output port.
3. **Delivery notification:** for each delivered cell, inform the corresponding input of the internal buffer status; that is, change the status of the corresponding VOQ to be eligible.

Figure 7(a) illustrates a scheduling cycle consisting of the three above-mentioned phases. The input scheduling is performed first, followed by the output scheduling, and finally the delivery notifying. It is of note that for fast implementation purposes, the output scheduling and the delivery notifying steps can be overlapped, as shown in Figure 7(b).

Features of the VOQ/BCS Architecture

This section presents the features that the VOQ/BCS offers. We will present these characteristics in terms of comparison with the input queued and the shared memory

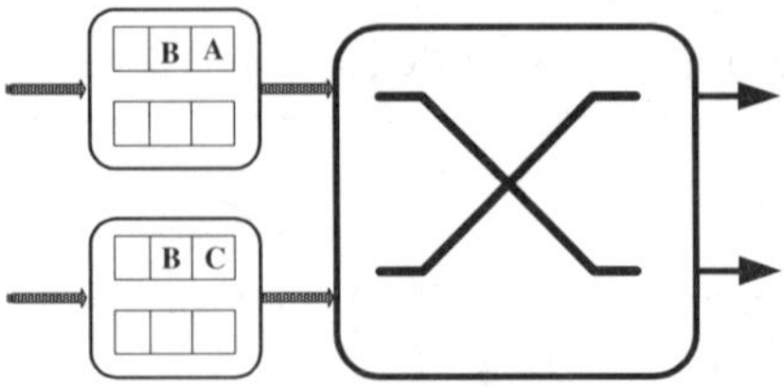

(a) 2×2 IQ crossbar switch

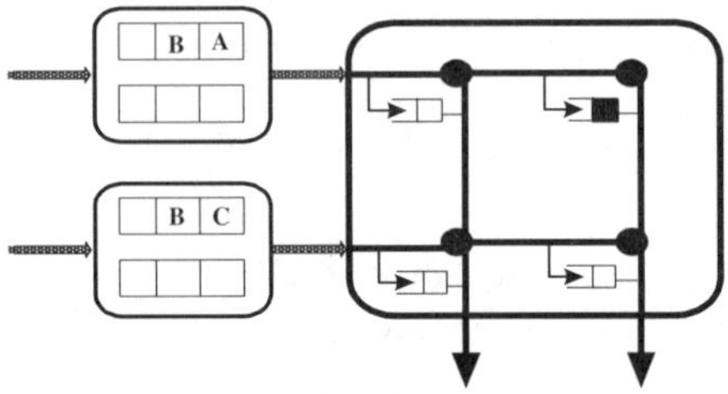

(b) 2×2 buffered crossbar switch

Figure 1.8. Architecture comparison of bufferless and buffered crossbar

architecture. Compared to the existing architectures, the VOQ/BCS architecture has substantial advantages. They can be summarized as follows:

- **Better performance:** In the IQ/VOQ architecture, a scheduling algorithm makes a match between inputs and outputs. An input can select an output provided that this output is not occupied by other inputs. This matching mechanism is usually done through the request–grant–accept phases. As was shown in [34], the decision time spent by the grant and the accept arbiters takes approximately 75% of the whole iteration time. This is due to high information exchange between the input and output schedulers, which enforces a high dependency between them, hence making their design complex and centralized. While in the VOQ/BCS architecture, if an output is not ready to receive the cell, the input can still transmit it to an internal buffer provided the internal buffer is empty. In other words, the choices of inputs and the choices of outputs needn't be synchronous. This can entirely resolve the problem of output contention, since an input can send to more than one output (via the internal buffer) and likewise, more than one output can receive cells coming from the same input. Figure 8(b) shows an example of output contention resolution. Adopting the VOQ/BCS architecture, the input schedulers and the output schedulers are totally distributed. There is no information exchange among them at all. This in turn, reduces the queuing delay and increases the throughput, hence improving dramatically the overall performance of the switch.

  The example shows a 2×2 IQ switch, Figure 8(a), and a buffered crossbar switch, Figure 8(b). In the first case, Figure 8(a), the two HoL cells (A and C) have

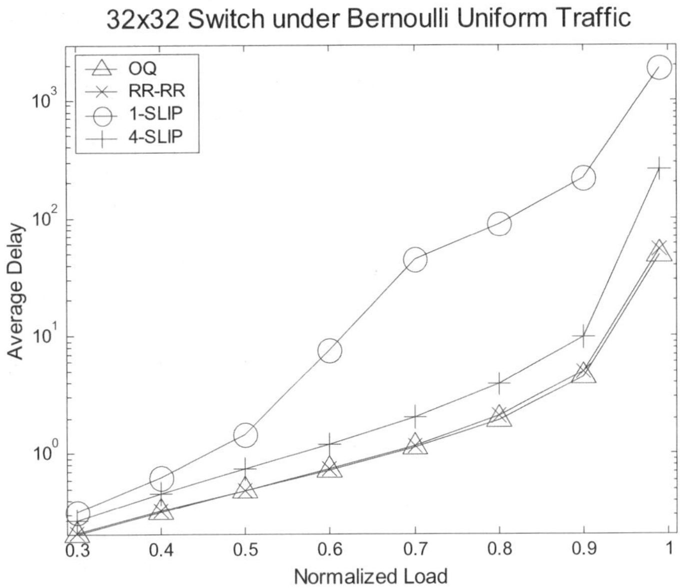

Figure 1.9. Delay performance under Bernoulli IID uniform traffic

the same destination output, and one of them can be forwarded because of the output contention. However, in Figure 8(b), the same input HoL cells A and C can be transmitted, simultaneously, to the internal buffers, since only $IB_{1,2}$ is full. Therefore the output contention problem does not occur at all for the buffered crossbar case, so long as the internal buffer is free.

To show the far better performance a VOQ/BCS exhibits when compared to a bufferless architecture, we plot the performance study of [28]. A 32×32 buffered crossbar switch, using a simple round-robin arbitration scheme, was simulated under Bernoulli IID uniform traffic and compared to that of a bufferless architecture using iSLIP with one and four iterations, as in Figure 1.9.

- **Higher speed and simpler implementation:** Many scheduling algorithms for the IQ/VOQ architecture are iterative (*e.g.* iSLIP, FIRM). However, the performance of a single iteration is often not good enough. For iSLIP, usually four iterations are employed. These additional iterations result in additional complexity. For this architecture, we just run the above three steps. Input scheduling and output scheduling can be totally independent, reducing largely their arbitration complexity, which is linear on the input ports number N.
- **Lower hardware requirement:** With shared memory, a speedup of N is required to transfer all the incoming cells simultaneously to their outgoing ports. However, for the VOQ/BCS architecture, each internal buffer is separated from one another. An internal buffer will be accessed at most two times: once by an input and once by an output.

However, for a large number of inputs/outputs, the number of internal buffers becomes huge (since it equals N^2). Even though VLSI density increases make it possible to embed enough memory on chip, it can still be quite difficult to put so much memory on a crossbar. One way to make the implementation easier is by moving the internal buffers out of the crossbar and putting them at the input side of the chip and adding two multiplexers. The first multiplexer at each input chooses one among the N VOQs and sends the HoL packet to the internal buffers, and the second multiplexer chooses one internal buffer according to the packet's destination output in which to put the received packet. Another implementation variant is putting the internal buffers at the output side instead of the input. The multiplexer at an output chooses one internal buffer to receive a packet and then sends it out. It is quite clear that the two variants are equivalent to the original architecture. We believe that these two alternative layouts are less complicated and more scalable than the conventional one.

Scheduling Cells in the VOQ/BCS Architecture

As bufferless scheduling algorithms reached their practical limitations due to higher port numbers and data rates, buffered crossbars received increased interest because they have the potential to solve the complexity and scalability issues faced by their bufferless predecessors. The increasing need for terabit switches and routers means that future commercial packet switches must be implemented with reduced scheduling complexity, and buffered crossbar architectures, inherently, can implement distributed scheduling schemes.

In the past few years, many scheduling schemes have been proposed for the VOQ/BCS architecture. The first, [3], used input and output arbitration schemes based on the *Oldest Cell First*. The second scheme presented was based on *round-robin* arbitration in both input and output scheduling [28]. While these two schemes were proven to achieve 100% throughput under uniform traffic, they performed rather poorly under non-uniform traffic patterns. To overcome this, a scheme based on *Longest Queue First* (LQF) input scheduling followed by a round-robin arbitration scheme in the output side was proposed and demonstrated 100% throughput under uniform traffic patterns.

It is of note that the algorithms presented above were compared with bufferless schemes and were demonstrated, as expected, to have much better performance. However, as mentioned earlier, the VOQ/BCS has key advantages that ensure the scheduling algorithm is simple and efficient at the same time. So far, most existing algorithms (OCF, RR, and LQF) are just simple mappings of previously proposed algorithms for bufferless crossbar switches into the new VOQ/BCS architecture.

The presence of internal buffers significantly improves the overall performance of the switch due to the advantages it offers. A scheme that takes full advantage of the internal buffers was recently presented in [35]. This scheme was, in fact, an approximation of the *First Come First Served* (FCFS) policy. It was based on the *Currently Arrival First* (CAF) cell in the input scheduling followed by a priority

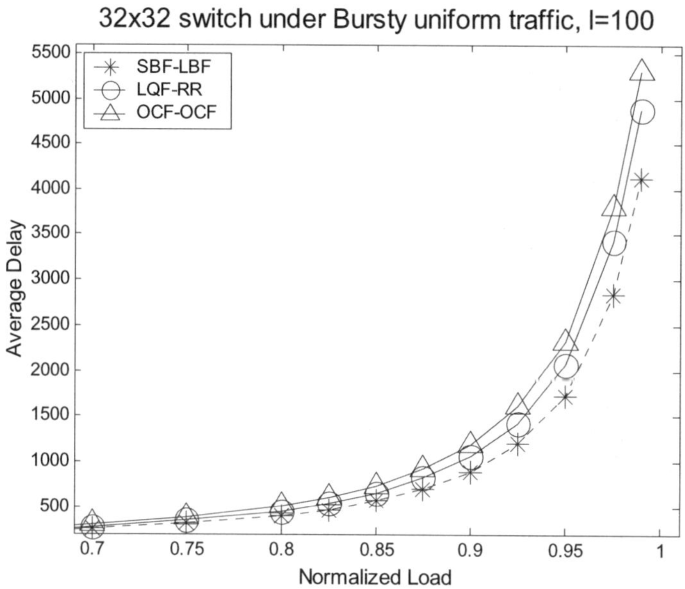

Figure 1.10. Delay performance under bursty uniform traffic, l =100

scheme called *Priority ReMoVal* (PRMV). A scheme that exclusively used the internal buffers was also proposed [36]. This scheme is called the *Most Critical internal Buffer First* (MCBF), and is based on the *Shortest internal Buffer First* (SBF) at the input scheduling and on the *Longest internal Buffer First* (LBF) at the output side. The authors addressed the important role that the internal buffer element plays in the scheduling process due to its shared nature. While being a stateless scheme, MCBF outperforms weighted scheduling schemes such as LQF and OCF. Figure 1.10 shows the delay performance of MCBF, LQF-RR and OCF-OCF under uniform burst traffic with burst length equal to 100.

1.4 Multi-stage Switching

1.4.1 Architecture Choice

Clos-network switch architectures can be categorized into two types. The first type has buffers in the second stage, such as the WUGS architecture in [6]. The function of the buffers is to resolve contention among cells from different first-stage modules. However, cells may be mis-sequenced at the output ports, requiring a re-sequence function, which is difficult to implement when the port speed increases. The second type of architecture has no buffers in the second stage and uses shared memory modules in the first and last stages to aggregate cells. The ATLANTA switch with its Memory/Space/Memory (MSM) architecture is a commercially successful example [7]. This architecture is more promising as no mis-sequence problem exists. We describe several dispatching algorithms for the MSM architecture, including concurrent dispatching (CD), concurrent round-robin dispatching (CRRD), concurrent master–slave round-robin dispatching (CMSD), and concurrent static round-robin dispatching (SRRD).

However, one disadvantage of the MSM architecture is that the input and output stages are both composed of shared-memory modules, introducing a memory speedup problem. Although the speedup is smaller than that in output-queued switches, it definitely hinders the switch's ability to scale up to very large port numbers.

The memory speedup problem was solved by the bufferless Clos-network architecture proposed in [37]. This architecture contains only crossbar switching elements in all stages. All cells are stored in the input port cards, as done with the virtual output queuing structure in single-stage crossbar switches. Since the switching elements are fully distributed by smaller modules, this raises the challenge of how to design the scheduling algorithm in a fully distributed way. We then describe the Distro dispatching algorithm for the bufferless Clos-network architecture.

1.4.2 The MSM Clos-network Architecture

Switch Model

The switch architecture used in this paper is based on [37] and is shown in Figure 1.11. The input and output stages are both composed of shared-memory modules, each with n port interfaces. They are fully interconnected through a central stage that consists of bufferless crossbars of size $k \times k$. In the switch, there are k input modules (IM), m central modules (CM), and k output modules (OM).

An OM(j) has n buffered output ports, OP(j,h). Each output port buffer can receive at most m cells from m central modules and send at most one cell to the output line in one time slot.

An IM(i) has nk virtual output queues, VOQ(i, j, h), for storing cells that go from IM(i) to OP(j, h) at OM(j). Each virtual output queue can receive at most n cells from n input ports and send one cell to the central module. We use VOQ Group (i, j) to represent all VOQs storing cells from IM(i) to OM(j).

An IM(i) has m output links, LI(i, r), connecting to each CM(r). An CM(r) has k output links, LC(r, j), connecting to each OM(j).

Concurrent Dispatching (CD)

The distributed architecture implies the presence of multiple contention points in the switch. The ATLANTA switch proposed the CD algorithm with highly distributed nature [7, 38]. It works as follows.

In each time slot, each IM randomly selects up to m VOQs and randomly sends the requests to CMs. If there is more than one request for the same output link in

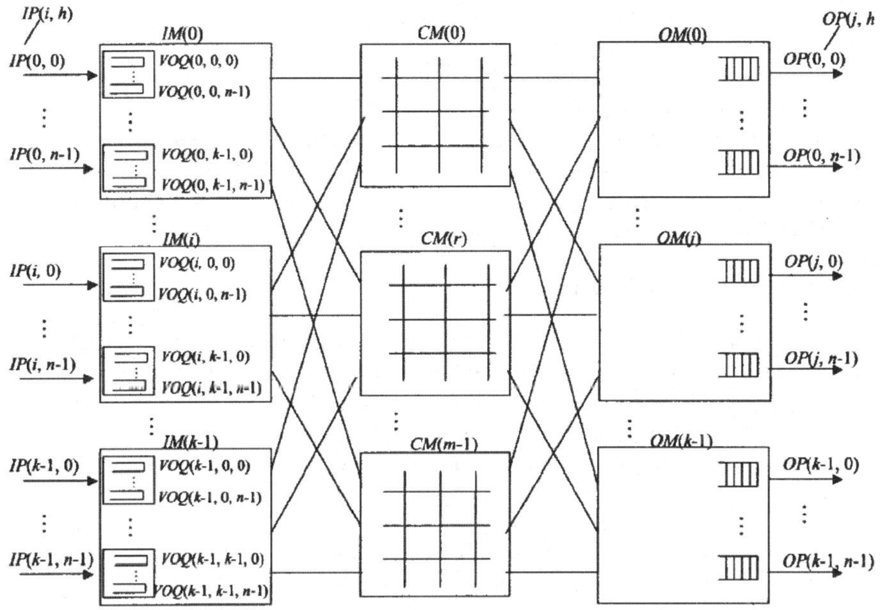

Figure 1.11. The MSM Clos-Network model

a CM, it grants one request randomly. Finally, the granted VOQs send to the corresponding OP in the next time slot.

However, the CD algorithm cannot achieve a high throughput unless the internal bandwidth is expanded. The original CD algorithm applies a backpressure mechanism in the dispatching process. We describe only its basic concept and characteristic in this paper. We also assume that the buffer size in the IMs and OMs is large enough to avoid cell loss. Hence we can focus the discussion on the properties of the dispatching algorithms.

Concurrent Round-robin Dispatching (CRRD)

In crossbar switches, round-robin arbitration has been developed to overcome the throughput limitation of the PIM algorithm. Similarly, the CRRD algorithm has been proposed in [39] to overcome the throughput limitation of the CD algorithm by using round-robin arbiters. It is based on the request-grant-accept (RGA) handshaking scheme, just as in the iSLIP algorithm. Since contention occurs in both output links of the IMs and CMs, two phases must be employed to resolve the contentions. In Phase 1, the algorithm employs iterative matching between VOQs and output links in each IM. Phase 2 then performs contention control for the output links in the CMs.

Phase 1: Iteractively Matching within IM:

- *Step 1 Request*: Each unmatched, non-empty VOQ(i, j, h) sends a request to every output link LI(i, r).
- *Step 2 Grant*: Each output link LI(i, r) selects one request with the round-robin arbiter and sends the grant to the selected VOQ.
- *Step 3 Accept*: Each VOQ V(i, j, h) selects one grant with the round-robin arbiter and sends an acceptance notice to the selected output link LI(i, r).

Phase 2: Matching between IM and CM:

- *Step 1 Request*: Each IM output link LI(i, r), which was accepted by VOQ(i, j, h) in Phase 1, sends the request to the CM output link LC(r, j).
- *Step 2 Grant*: Each CM output link selects one request with the round-robin arbiter. It then sends the grant to the selected IM.
- *Step 3 Accept*: If the IM receives the grant from the CM, it sends the head cell from the matched VOQs in the next time slot.

Note that CRRD uses three sets of round-robin arbiters to resolve the contentions in IMs and CMs. As with iSLIP, the round-robin pointer in the arbiters is updated to one position after the selected position if and only if the match within the IM is achieved in Phase 1 and the request is granted by the CM in Phase 2. The desynchronization effect of the round-robin pointers in CRRD works exactly as in iSLIP. As a result, CRRD provides 100% throughput under uniform traffic and burst traffic independent of the number of iterations in Phase 1.

Concurrent Master/Slave Round-robin Dispatching (CMSD)

The CMSD is an improved version of the CRRD. It employs two sets of arbiters in the IM, a master and a slave. They operate concurrently in a hierarchal round-robin manner. The CMSD differs from the CRRD only in the iterative matching process in Phase 1.

Phase 1: Iteractively Matching within IM:

- *Step 1 Request*: Each unmatched, non-empty VOQ(i, j, h) sends a request to the jth slave arbiters in every output link arbiter LI(i, r). At same time, each VOQ Group (i, j) that has at least one unmatched, non-empty VOQ sends a request to the master arbiter in every output link LI(i, r).
- *Step 2 Grant*: Each slave arbiter and master arbiter simultaneously selects the request in a round-robin fashion. At same time, each master arbiter selects one VOQ Group's request in a round-robin fashion. Finally, LI(i, r) sends the grant to VOQ(i, j, h) if and only if j has been selected by the master arbiter and h has been selected the jth slave arbiter.
- *Step 3 Accept*: Each VOQ(i, j, h) selects one grant in a round-robin fashion and sends an acceptance notice to the selected output link LI(i, r).

CMSD provides enhanced scalability while preserving the advantages of CRRD. In CRRD, each arbiter in LI(i, r) should make the decision among nk requests. In CMSD, master arbiters need only consider k requests and slave arbiters n requests. Since all arbiters can operate simultaneously, this will considerably reduce the arbitration time.

Concurrent Static Round-robin Dispatching (SRRD)

The Static Round-robin (SRR) scheme has been demonstrated to give very good delay performance in crossbar switches. The key idea is to desynchronize the pointers in the round-robin arbiters in a static way and to use a rotating-search technique to improve the performance under non-uniform traffic. Intuitively, we can apply the SRR technique into dispatching processes in the Clos-network switching system. The Static Round-robin Dispatching (SRRD) scheme was proposed in [40].

SRRD is exactly the same as CMSD except in its method for updating the round-robin pointers. The novelty of our SRRD scheme is the systematic initialization and intelligent updating of the pointers. All pointers are artificially set to be desynchronized to efficiently resolve the contentions in each time slot, and it is guaranteed that no VOQ will be starved in every N time slots.

Performance Evaluation

We compare the delay performance of the dispatching algorithms in the MSM architecture and in the single-stage crossbar switch architecture under uniform traffic. We use the Clos-network setting $m = n = k = 8$, which corresponds to a port size of $N - 64$ in the crossbar switch.

As shown in Figure 1.12, the average delay of the algorithms in the MSM architecture is larger than those in the crossbar switch when load is below 0.5. This is because most of the dispatching algorithms in the crossbar switch perform fairly efficiently under low traffic load. The MSM architecture introduces more contention points in the distributed modules, however, the performance of the architecture is improved significantly in the heavy load region, as shared memory modules are effectively used.

The randomness-base algorithms, PIM and CD, could only achieve about 60% throughput. All remaining round-robin algorithms can achieve 100% throughput under uniform traffic. SRRD has the lowest average delay and is much closer to the performance of an output-queued switch. This result shows the significant effect of the SRR technique in the MSM architecture.

1.4.3 The Bufferless Clos-network Architecture

Switch Model

One disadvantage of the MSM architecture is that the input and output stages are both composed of shared-memory modules. This results in a memory speedup of n

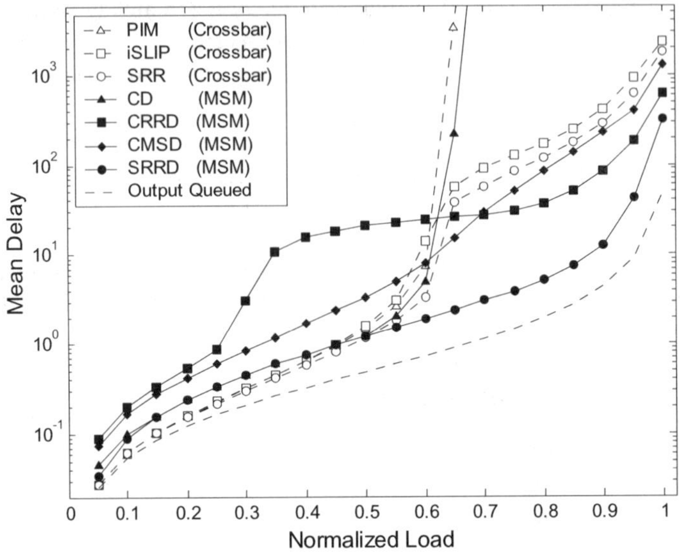

Figure 1.12. Delay comparison of crossbar and MSM

in each IM and k in each OM. In output-queued switches, the memory speedup is $N = n \times k$. Although the speedup of MSM is smaller than that in output-queued switches, it definitely hinders a switch's ability to scale up to very large port numbers. In [37], a bufferless Clos-network switching architecture has been proposed.

As depicted in Figure 1.13, the bufferless Clos-network architecture is slightly modified from the MSM architecture by replacing all shared memory modules by crossbars. All cells are stored in the input port cards, the same as the virtual output queuing structure in single-stage crossbar switches.

An IP(i, g) has N virtual output queues, VOQ(i, g, j, h), storing cells that go from IP(i, g) to OP(j, h) at OM(j). Each virtual output queue can receive at most one cell and send at most one cell. A VOQ Group (i, g, j) comprises all VOQs from IP(i, g) to OM(j). In this paper, i corresponds to an IM, g to a specific input port of an IM, j corresponds to an OM, and h to a specific output port of an OM.

Distro Dispatching Algorithm

Since contention points exist in all output links of the IPs, IMs, CMs and OMs, scheduling in the bufferless architecture is more challenging than in the MSM architecture. The Distributed Static Round-robin (Distro) dispatching algorithm has been proposed in [37] to dispatching cells in the new architecture. The Distro algorithm is a natural extension from algorithms in the MSM architecture.

The additional two phases are included in Distro because elimination of shared memory modules in the first and third stage introduces two more contention points.

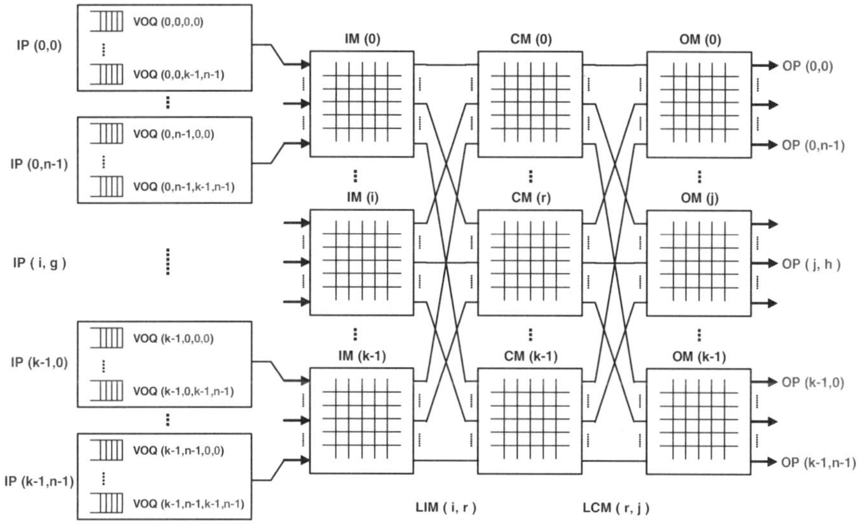

Figure 1.13. The bufferless Clos-network model

Note that the algorithms for the MSM architecture are based on the request-grant-accept (RGA) handshaking scheme. This approach is difficult to implement when too many contention points exist, and therefore the Distro algorithm adopts the request-grant (RG) scheme as proposed in the DRRM algorithm. Similar to SRRD, the Distro algorithm systematically initializes and intelligently updates the pointers. As a result, all pointers are artificially set to be desynchronized to efficiently resolve the contentions in each time slot, and it is guaranteed that no VOQ will be starved in every N time slots.

Each IP(i, g) is associated with Arbiter_j (i, g), VOQ Group in IP(i, g) with Arbiter_h (i, g, j), LI(i, r) with Arbiter_g (i, r), LC(r, j) with Arbiter_i (r, j), and OP(j, h) with Arbiter_r (j, h). All arbiters are making decisions in a round-robin fashion. Round-robin pointers of the highest priority are desynchronized at the start and will be updated systematically to keep the desynchronization in each time slot. The Distro algorithm works as follows:

- **Phase 1 Request selection in each IP(*i,g*):** Each Arbiter_j (i, g) selects an non-empty VOQ Group (i, g, j). At the same time, each Arbiter_h (i, g, j) selects a non-empty VOQ(i, g, j, h) within VOQ Group (i, g, j). Then each IP(i, g) sends the request $[j, h]$ to its output link.
- **Phase 2 Grant from LI(*i,r*):** Each LI(i, r) systematically chooses the request $[j, h]$ from IP(i, g) and sends the request to LC(r, j).
- **Phase 3 Grant from LC(*r,j*):** If LC(r, j) receives one or more non-empty requests from k LIs, it chooses the request $[j, h]$ in LI(i, r) with Arbiter_i (r, j) and sends the request to OP(j, h).
- **Phase 4 Grant from OP(*j,h*):** If OP(j, h) receives one or more non-empty requests from k LCs, it chooses the request $[j, h]$ in LC(r, j) with Arbiter_r (j, h).

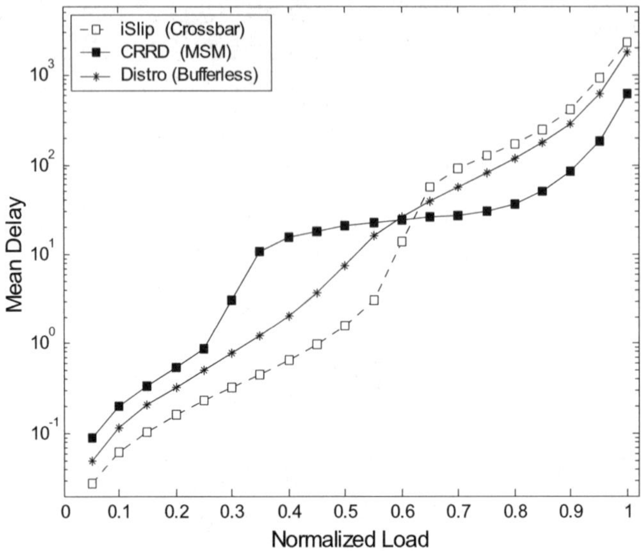

Figure 1.14. Delay comparison with crossbar and multi-stage scheduling algorithms

Finally, OP(j, h) notifies the IP(i, g) via the granted path, and the VOQ (i, g, j, h) will send to OP(j, h) in the next time slot.

As mentioned before, the bufferless Clos-network architecture is very scalable for port size. There are actually many flexibilities in the configurations. We can either scale up the port size by increasing the number of ports n per input/output module, or increasing the number of central modules m. Note that m must be larger than or equal to n in order to achieve the non-blocking property in Clos networks.

Performance Evaluation

We compare the delay performance of our Distro algorithm with other related algorithms in Figure 1.14. It is clear that Distro achieves 100% throughput under uniform traffic. When load is less than 0.65, the Distro algorithm is worse than iSLIP on a reasonable scale. This is due to an increased number of contention points in the Clos-network switches. However, as the load increases, the desynchronization effect of Distro improves the delay performance. In the heavy load region, the performance of Distro closely approximates to the performance of the SRR algorithm.

The delay performance of the MSM algorithms is generally worse than other algorithms in the light load region. Since the MSM algorithms use shared-memory modules to resolve the contention for OPs, their delay performance in the heavy load region are the best when compared with other architectures. This is compensated by the high memory speedup.

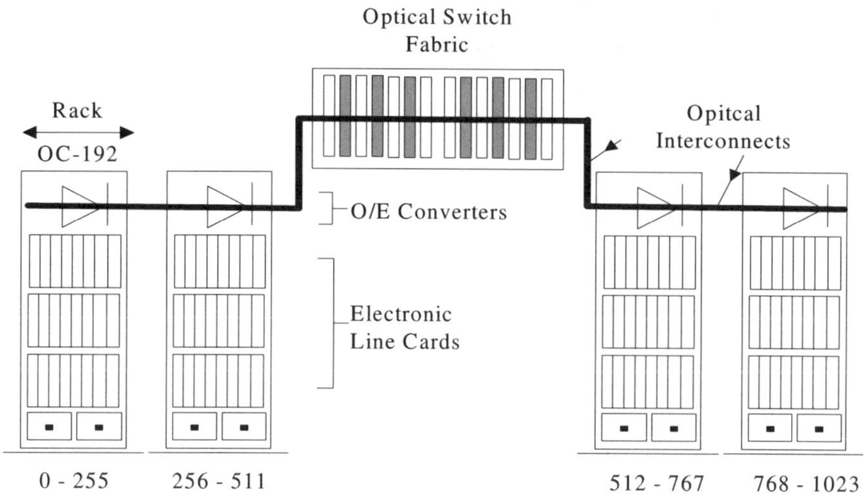

Figure 1.15. Illustration of a multi-rack switching system (1024 × 1024, 4 racks)

1.5 Optical Packet Switching

1.5.1 Multi-rack Hybrid Opto-electronic Switch Architecture

The explosion of Internet traffic has brought about an acute need for broadband communication networks. This heightens the demand for high-performance switches / routers more than ever and terabit switching systems are now receiving much attention. However, such systems with high power consumption and large numbers of ports can no longer be built in a compact, single-rack fashion [41]. Most high-capacity switches currently under development employ a multi rack system architecture, as shown in Figure 1.15, with the switching fabric in one rack and line cards spread over several racks. Such racks are connected to each other via cables.

Besides the overall architecture there is the choice of hardware switch components. This includes:

- *Packet buffering and processing*: because of the immaturity of optical buffering technology, such as fiber delay lines (FDL) [42], packets are still buffered and processed electronically.
- *Interconnects*: cables replace the backplane in multi-rack systems. Interconnects can be copper or parallel optical cables. However, compared to copper cables, parallel optics are favored for higher bandwidth, lower latency and extended link length beyond copper capabilities [43].
- *Fabric*: The fabric can be electronic or optical. Suppose a traditional electronic fabric is used. In such a system architecture, a series of opto-electronic conversions are needed to switch packets. Line cards terminate high-capacity optical fiber links from the network, and the received packets are processed and buffered electronically. Since the switch is located in an independent rack, fiber

links are used to transfer packets between the line cards and the switch fabric. It is clear from the above description that the system introduces an additional layer of opto-electronic (O/E) conversions between the interconnect and the switch fabric, which results in high complexity and significantly higher cost. In addition, the central electronic fabric occupies valuable floor space, consumes a lot of power, and poses several reliability problems owing to the large number of high-speed active components.

An obvious evolution to this architecture is to try to replace the central electronic fabric with an optical fabric [44, 45, 46, 47, 48]. Optical switch fabrics may overcome the cost, power, space, and scalability problems that arise in sizing traditional electrical backplanes into the terabit regime.

To this end, an opto-electronic hybrid switch based on the strengths of both electronics and optics is a practical and viable approach. This architecture uses electronics for packet processing/scheduling/buffering and optics for packet transmission and switching.

1.5.2 Optical Fabrics

The basic premise of optical switch fabrics is their high switching capacities and reduction in O/E conversions. These advantages are significant as there is no need for lots of expensive high-speed electronics. Furthermore, optical fabrics are cheaper and smaller in physical size. They may also provide potential benefits including scalability, high bit rate, and low power consumption. The technologies include optical micro-electromechanical systems (MEMS)-based switching [49][50], thermal optical switching [51], electro-optic switching, bubble switches [52], etc. Reference [53] gives a detailed comparison between these optical switching technologies.

We use optical MEMS fabric as an example. The basic switching elements of MEMS are tiny micro-actuated free-rotating mirrors as shown in Figure 1.16. The switching function is performed by reflection of the light. MEMS has several hundred of these tiny mirrors arranged in arrays and integrated on a silicon chip. The two-dimensional (2D) optical MEMS has mirrors arranged in a crossbar configuration [54]. Figure 1.17 shows a photo of a 2D $N \times N$ switch fabric made by AT&T [55]. Collimated light beams propagate parallel to the substrate plane. When a mirror is activated, it moves into the path of the beam and directs the light to one of the outputs by making a 45° angle with the beam. In general, the (i, j) mirror is raised to direct light from the ith input fiber to the jth output fiber. The mirror is no larger in diameter than a human hair. Several hundred such mirrors can be built on a chip no larger than a few centimeters square. Since MEMS create so many mirrors on a single chip, the cost per switching element is relatively low. Therefore, MEMS allow low-loss large-port-count optical switching solutions at very low cost per port.

Although there are many advantages to using optical fabrics as mentioned before, these technologies are still emerging and usually exist in sub-optimal form today. The reconfiguration times (or reconfiguration delay) of optical fabrics are much longer than those of electronic fabrics. For example, a MEMS-based optical fabric needs

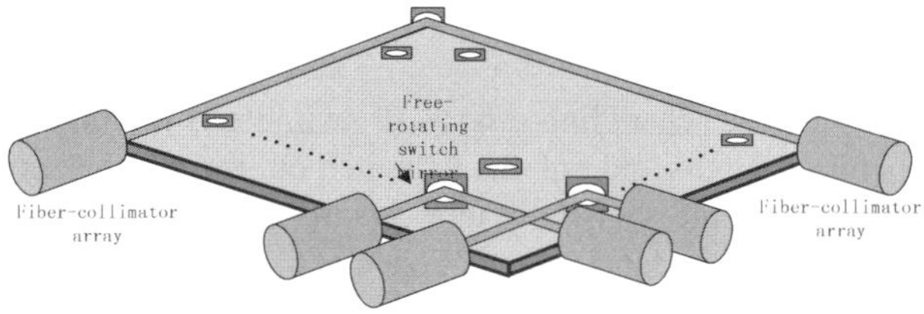

Figure 1.16. Illustration of a 2D MEMS approach to construction of the optical switch fabric

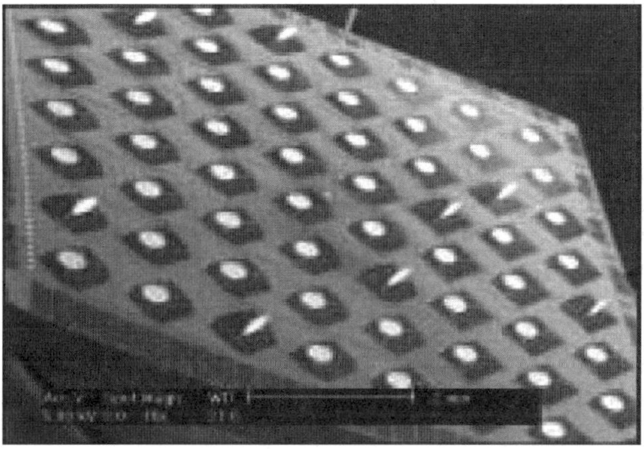

Figure 1.17. 2D $N \times N$ switch fabric demonstrated by AT&T

Table 1.1. Reconfiguration delay times for some optical switching technologies

Switching technology		Delay
Optical MEMS	Mirror/gap-closing electrostatic actuator	7 ms
	1 × 2 MOEMS switch based on silicon-on-insulator and polymeric waveguides	32–200 ns
Thermal optical switch	Bubble-actuated switch	1 ms
	Fully packaged polymeric four arrayed 2 × 2 DOS	< 5 ms
Electro-optic switch	Electroholographic (EH) optical switch (1 × 2)	< 10 ns
	Liquid crystal holographic optical switch (1 × 8)	ms
	Electronically switchable waveguide Bragg gratings switch (2 × 2)	10–50 ns

to adjust the angles of the fabric mirrors to set up new connections. This introduces mechanical settling, synchronization, and other time-consuming operations. These operations take times from hundreds of nanoseconds to a few milliseconds [53]. Table 1.1 samples the reconfiguration delay times for several different technologies. This is around 10 to 10^5 time slots for a system with a slot time equal to 50 ns (64 bytes at 10 Gb/s).

1.5.3 Reduced Rate Scheduling

As introduced earlier, the scheduling task in all switches is to find a one-to-one bipartite match between inputs and outputs to switch out the packets, no matter if the fabric is optical or electronic. Generally in each time slot, the electronic switch runs the scheduling, sets up a new fabric connection, and then transfers the corresponding cells.

However, this slot-by-slot approach is not practical for optical switches. For example, if the reconfiguration delay equals one time slot, one cell transmission takes two time slots. One time slot is for scheduling and connection setup (fabric reconfiguration), the second time slot is for actual transmission. In other words, only half of the fabric bandwidth is used for cell transmission (which is effective bandwidth); while the other half is wasted for reconfiguration. For a switch which transfers 64B packets at a 40 Gb/s line rate and suffers a 100 ns reconfiguration delay, the effective bandwidth is as low as 10%. If the reconfiguration delay is not addressed by a proper scheduling scheme, it can severely cripple the performance of optical switches. It is clear that the scheduling rate must be reduced to compensate for the reconfiguration delay. Each schedule holds for some time rather than changing at every time slot.

The reduced rate scheduling is not a simple extension of scheduling algorithms introduced in Section 1.2. It has been proven that the scheduling problem for optical switches with reconfiguration delay is NP-hard [48]. Currently, most of the reduced rate scheduling algorithms use the time slot assignment (TSA) approach [56, 57, 58, 48] and provide heuristic solutions.

1.5.4 Time Slot Assignment (TSA) Approach

The TSA-type of algorithm periodically accumulates incoming traffic and maps this batch to a set of switch configurations. The objective of TSA scheduling in an optical switch is to find a set of fabric configurations (or *schedules*) and their respective holding time, to switch out all of the accumulated traffic, and to maximize the fabric utilization. In other words, this is equivalent to minimizing the total *transmission makespan*, which includes the time spent in transmitting traffic and the time spent in reconfiguring the fabric. This is proved to be NP-hard under non-zero reconfiguration delay [48]. All algorithms introduced here are heuristic.

Scheduling Scheme

Using the TSA scheduling scheme, the switch works in a three-stage accumulation, scheduling, and transmission cycle. The length of the accumulating stage is set to be a predefined system constant T. Incoming traffic in these T time slots are stored in traffic matrix D. For a $N \times N$ switch, $D = (d_{ij})$ is a nonnegative integer $N \times N$ matrix. d_{ij} represents the number of cells received in input i whose destinations are output j during the accumulating stage. The scheduling stage then finds a set of N_s schedules $P_1, P_2, \ldots, P_{N_s}$ for the accumulated traffic. Each schedule will be held for $\phi_1, \phi_2, \ldots, \phi_{N_s}$ time slots respectively. Following the scheduling decision, the

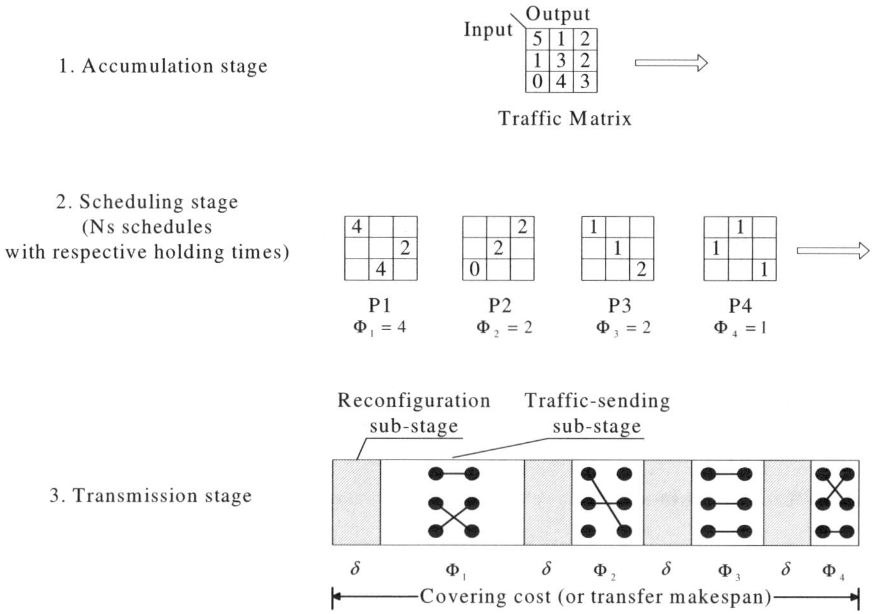

Figure 1.18. TSA scheduling example

transmission stage sets up the fabric and switches out the traffic repeatedly until all packets are sent out. The transmission makespan is equal to $\sum_{k=1}^{N_s} \phi_k + N_s\delta$. Figure 1.18 shows an example of TSA scheduling in a 3×3 switch.

The three-stage TSA operations can be further accelerated by using a pipelining framework as shown in Figure 1.19, with each stage running in T time slots. Notice that since the transmission stage suffers from a $N_s\delta$ reconfiguration delay and potentially empty time slots (that exist when a particular connection is held with no more packets to transmit), speedup S is generally needed to ensures it finishes within T time slots.

TSA algorithms are favored because of their good features:

- **Stability:** because the traffic batch gathered in the accumulation stage is always sent out in the transmission stage, TSA scheduling is stable under any admissible traffic patterns.
- **Bounded cell delay:** the pipeline diagram indicates that a cell will go through the switch within $3T$ slot times. This bounded worst cell delay ($3T$) makes it possible to provide QoS guarantees.

For the required buffer size, it is observed that traffic from at most three different batches may appear in a particular input buffer at the same time. Assume B is the number of bits sent to one input per time slot. A buffer of size $3TB$ is enough for each input and $2TB$ for each output. If all ports are considered, the switch needs at most $5TBN$ bits buffer size.

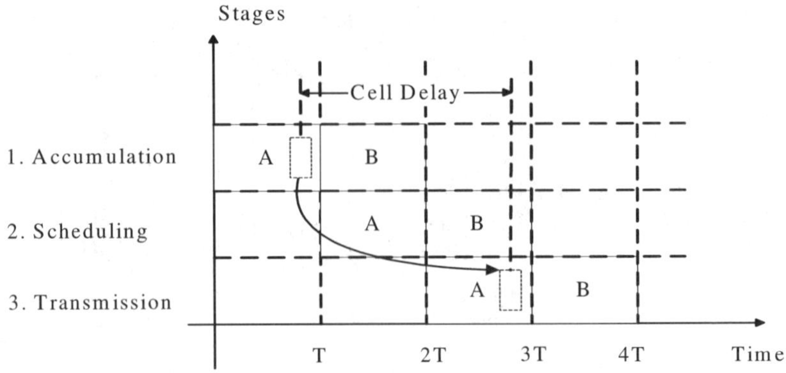

Figure 1.19. Pipelined TSA execution with two traffic batches

1.5.5 DOUBLE Algorithm

Early TSA algorithms assume the reconfiguration delay to be zero [56] or infinite [57] for simplicity. However, the extreme assumptions on reconfiguration delay are no longer practical in optical switching systems. A recent DOUBLE algorithm [58] considers scheduling under moderate reconfiguration delays. It performs better than algorithms with extreme assumptions under a large spectrum of reconfiguration delay values [58].

DOUBLE works by separating the traffic matrix D into *coarse* and *fine* matrices and devotes N configurations to each. Coarse matrix A and fine matrix B are generated in a way that ensures $D \leq \lceil T/N \rceil A + B$. The algorithm first generates the coarse matrix A by dividing the elements of D by T/N and taking the floor. The rows and columns of A sum to at most N (because of the admissible traffic assumption), thus the corresponding bipartite multigraph can be edge-colored in N colors. Each subset of edges assigned to a particular color forms a match, which is weighted by $\lceil T/N \rceil$. The fine matrix B for D does not need to be explicitly computed because its elements are guaranteed to be less than $\lceil T/N \rceil$. Thus, any N configurations that collectively represent every entry of B, each weighted by $\lceil T/N \rceil$, can be used to cover the fine portion. An example of DOUBLE execution is shown in Figure 1.20.

In summary, DOUBLE generates $2N$ schedules, each with holding length $\lceil T/N \rceil$. The total transmission makespan is $2N \times \lceil T/N \rceil + 2N \times \delta = 2T + 2N\delta$. DOUBLE has a time complexity of $O(N^2 \log N)$, which is mainly determined by the edge-coloring algorithm [59].

1.5.6 ADJUST Algorithm

Although the DOUBLE algorithm greatly enhances scheduling performance, its scheduling decision is rigid and does not change according to different system parameters. For example, if there are two switching systems with the same switch port number and accumulation time, but with different reconfiguration delay, DOUBLE

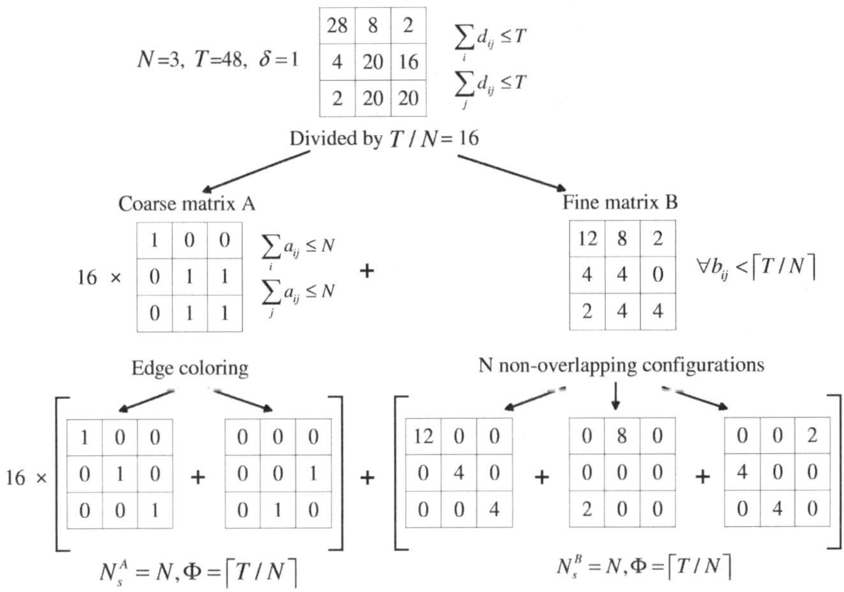

Figure 1.20. Example of the DOUBLE algorithm

will generate identical scheduling. However, it is natural that when the reconfiguration delay increases, the number of schedules should decrease to avoid incurring large overhead. Furthermore, simulation shows that a fixed choice of matrix division factor (T/N) may cause a large fine matrix with high cost, especially when the reconfiguration delay is relatively small compared to the accumulation length. All the above greatly influence the flexibility and performance of the DOUBLE algorithm. An enhancement ADJUST algorithm is therefore proposed. By introducing a regulating factor $\lambda = \sqrt{T/\delta N}$, it is able to self-adjust with different systems.

Similar to DOUBLE, ADJUST works by first separating the traffic D into a *quotient* matrix and a *residue* matrix and assigning schedules to each. The matrices are generated by dividing the elements in D by $T/\lambda N$. Under such a division, quotient and residue matrices can be covered by λN and N configurations, respectively. All configurations hold for $\lceil T/\lambda N \rceil$ time slots. The total transmission makespan is then $(T + \delta N) + (\delta \lambda N + \frac{T}{\lambda})$ (refer to [48] for details). The makespan is minimized when $\lambda = \sqrt{T/\delta N}$. An example of ADJUST execution is shown in Figure 1.21. For $N = 3$, $T = 48$ and $\delta = 1$, the regulating factor $\lambda = \sqrt{T/\delta N} = 4$.

It is obvious that the ADJUST algorithm always outperforms DOUBLE in the sense of makespan minimization. In fact, the DOUBLE algorithm can be viewed as a special case with λ set to 1.

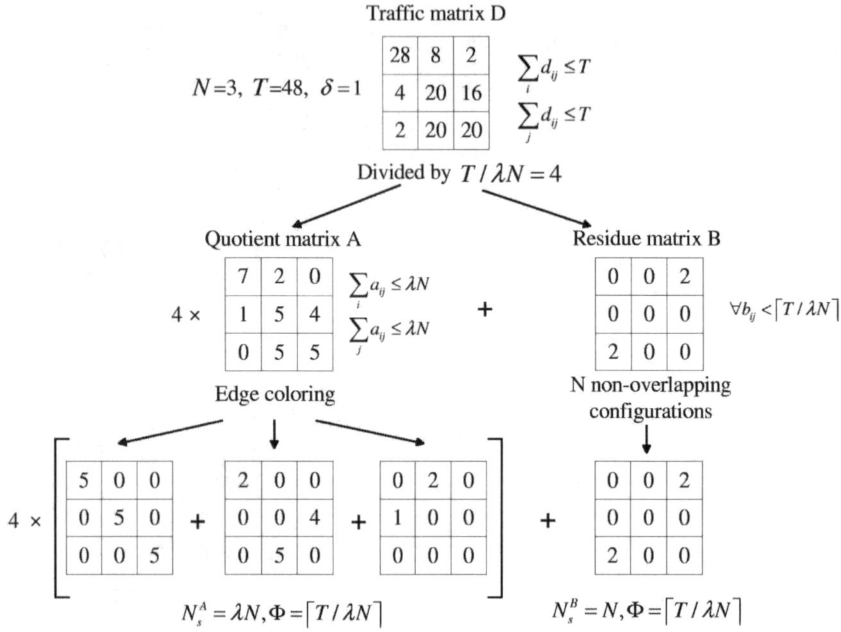

Figure 1.21. Example of the ADJUST algorithm

1.6 Conclusion

In this chapter, we have presented and discussed different router and ATM switch designs. We have discussed the bufferless switch architecture, the buffered switch architecture, the Clos-network architecture and the optical switching architecture.

The IQ crossbar switch is of interest due to its low hardware requirements, scalability and popularity. An IQ switch adopting a FIFO queue structure suffers from the HoL problem, which reduces the throughput of the switch to nearly half its available bandwidth. The speedup-based solution relies on the availability of advanced technology in terms of fast memory. However, this is totally impractical for high-performance routers. Very high speed is either uneconomical or simply impossible to achieve. The well-known efficient VOQ architecture is used to overcome the blocking problem and is of interest because of its ability to improve the switching capacity by several orders of magnitude. Despite its popularity, the VOQ solution has its own problems. A centralized scheduler is required to configure the switch so that, at each time slot, each input sends one cell to at most one output and each output receives one cell from at most one input.

In order to alleviate the centralized scheduler's task, the VOQ/BCS was proposed. The VOQ/BCS architecture is based on the strength of the VOQ architecture at the input side coupled with the internally buffered crossbar. This architecture has many advantages compared to the IQ switch. The internal buffering element enables totally distributed VOQs arbitration and the scheduler is not centralized anymore. The distributed nature of the arbitration improves the switching performance by several orders of magnitude. By means of computer simulation, we showed the improvement of the VOQ/BCS over the IQ switch. Furthermore, the VOQ/BCS doesn't need any additional cost such as speedup.

In an attempt at scalability, single-stage switching techniques are considered to be inherently limited by their quadratic complexity. The Clos-network architecture is widely recognized as a very scalable architecture for high-speed switching systems. So far, only limited success has been reported in the design of practical distributed scheduling schemes for the Clos-network. The ATLANTA switch with MSM architecture is an example of a commercially successful Clos-network switch. However, it necessitates internal bandwidth speedup. Recently, the CRRD and CMSD algorithms have been proposed in order to overcome the throughput limitation and implementation complexity problem. Based on the static round-robin technique, SRRD has been proposed to reduce the delay performance and deduce the hardware complexity. However, due to the memory speedup up problem in the shared memory modules, the MSM arrangement of the Clos-network switches hinder the scalability to very large port size. A bufferless Clos-network architecture has been proposed in which the shared memory modules are replaced by bufferless crossbar switching components. Although it requires a greater communication overhead among distributed modules, the bufferless Clos-network architecture is an efficient way to solve the scalability problem in switching systems.

One attractive direction in building scalable switches and routers is to adopt optical switching techniques. However, a main challenge remains in efficiently queuing the packets in the optical domain. The techniques (*i.e.* fiber delay lines) are not yet mature enough to support such functionalities. A viable solution is to have a hybrid switch architecture, with electronic buffers and optical fabrics. Having optical fabrics provides more scalability and higher switching speed over the electronic counterparts.

We are still facing many challenging requirements pertaining to the design of scalable and high-performance routers. In summary, the new generation of switches and router should first be highly scalable in port number and interface speed. Moreover, routers have to offer many services to meet today's customer needs such as guaranteed delay, bounded delay variation, minimal packet loss and controlled access. Finally, a router has to be cost effective.

References

1. M. Karol, M. Hluchyj, and S. Morgan. Input versus output queuing on a space division switch. *IEEE Transaction on Communications*, 35(12):1347–1356, 1987.
2. N. McKeown. islip: A scheduling algorithm for input-queued switches. *IEEE/ACM Transaction On Networking*, 7(2):188–201, Apr 1999.
3. M. Nabeshima. Performance evaluation of combined input-and crosspoint-queued switch. *IEICE Transaction on Communicaitons*, E83-B(3), Mar 2000.
4. S. Nojima, E. Tsutsui, H. Fukuda, and M. Hashimmoto. Integrated packet network using bus matrix. *IEEE Journal on Selected Areas in Communications*, 5(8):1284–1291, Oct 1987.
5. L. Mhamdi, M. Hamdi, C. Kachris, S. Wong, and S. Vassiliadis. High-performance switching based on buffered crossbar fabrics. To appear in *Computer Networks Journal.*
6. T. Cheney, J.A. Fingerhurt, M. Flucke, and J.S. Turner. Design of a gigabit atm switch. *INFOCOM*, pages 2–11, Apr 1997.
7. F.M. Chiussi, J.G. Kneuer, and V.P. Kumar. Low-cost scalable switching solutions for broadband networking: the atlanta architecture and chipset. *IEEE Communication Magazine*, 35(3):44–53, Dec 1997.
8. A. Demers, S. Keshav, and S. Shenker. Analysis and simulation of a fair queuing algorithm. *Inernetworking: Research and Experience*, 1(1):3–26, Sep 1990.
9. H. Zhang. Service disciplines for guaranteed performance service in packet-switching networks. *Proceeding of the IEEE*, 83(10):1374–96, Oct 1995.
10. S. T. Chuang, A. Goel, N. Mckeown, and B. Prabhakar. Matching output queueing with a combined input output queued switch. *INFOCOM*, pages 1169–1178, Apr 1999.
11. S. Y. Li. Theory of periodic contention and its application to packet switching. *INFOCOM*, pages 320–325, 1998.
12. N. McKeown, V. Anantharam, and J. Walrand. Achieving 100% throughput in an input-queued switch. *INFOCOM*, pages 296–302, Mar 1996.
13. A. Mekkittikul and N. McKeown. A starvation-free algorithm for achieving 100% throughput in an input-queued switch. *ICCCN*, pages 226–231, Oct 1996.
14. A. Mekkittikul and N. McKeown. A practical scheduling algorithm to achieve 100% throughput in input-queued switches. *INFOCOM*, pages 792–799, Apr 1998.
15. T. Anderson, S. Owicki, J. Saxe, and C. Thacker. High speed switch scheduling for local area networks. *ACM Transaction on Computer Systems*, 11(4):319–352, Nov 1993.
16. D. N. Serpanos and P. I. Antoniadis. FIRM: A class of distributed scheduling algorithms for high-speed atm switches with multiple input queues. *INFOCOM*, pages 548–555, Mar 2000.
17. L. Tassiulas. Linear complexity algorithms for maximum throughput in radio networks and input queued switches. *INFOCOM*, 2:533–539, Mar 1998.
18. P. Giaccone, B. Prabhakar, and D. Shah. Towards simple, high-performance schedulers for high-aggregate bandwidth switches. *INFOCOM*, pages 1160–1169, Jun 2002.
19. D. Gale and L. S. Shapley. College admissions and the stability of marriage. *American Mathematical Monthly*, 69(9):9–14, 1962.
20. I. Stoica and H. Zhang. Exact emulation of an output queueing switch by a combined input output queueing switch. *IEEE/IFIP IWQoS*, pages 218–224, May 1998.
21. P. Krishna, N. S. Patel, A. Charny, and R. J. Simcoe. On the speedup required for work-conserving crossbar switches. *IEEE Journal on Selected Areas in Communications*, 17(6):1057–1066, Jun 1999.
22. R. Bakka and M. Dieudonne. Switching circuits for digital packet switching network. United States Patent 4,314,367, Feb 1982.

23. M. Katevenis. Fast switching and fair control of congested flow in broad-band networks. *IEEE Journal on Selected Areas in Communications*, 5(8):1315–1326, Oct 1987.
24. A. K. Gupta, L. O. Barbosa, and N. D. Gorganas. 16 $\times$ 16 limited intermediate buffer switch module for atm networks for b-isdn. *GLOBECOM*, pages 939–943, Dec 1991.
25. M. Lin and N. Mckeown. The throughput of a buffered crossbar switch. *IEEE Communications Letters*, 9(5):465–467, May 2005.
26. S. Chuang, S. Iyer, and N. McKeown. Practical algorithms for performance guarantees in buffered crossbars. *INFOCOM*, pages 981–991, Mar 2005.
27. F. Abel, C. Minkenberg, R. Luijten, M. Gusat, and I. Iliadis. A four-terabit packet switch supporting long round-trip times. *IEEE Micro Magazine*, 23(1):10–24, Jan/Feb 2003.
28. R. Rojas-Cessa, E. Oki, Z. Jing, and H. J. Chao. CIXB-1: Combined input one-cell-crosspoint buffered switch. *HPSR*, pages 324–329, May 2001.
29. R. Rojas-Cessa, E. Oki, Z. Jing, and H. J. Chao. CIXB-K: Combined input-crosspoint-output buffered packet switch. *GLOBECOM*, pages 2654–2660, Nov 2001.
30. D. Stephann and H. Zhang. Implementing distributed packet fair queuing in a scalable switch architecture. *INFOCOM*, pages 282–290, Apr 1998.
31. M. Katevenis and G. Passas. Variable-size multipacket segments in buffered crossbar (CICQ) architectures. *ICC*, pages 999–1004, May 2005.
32. L. Mhamdi and M. Hamdi. Output queued switch emulation by a one-cell-internally buffered crossbar switch. *GLOBECOM*, 7:3688–3693, Dec 2003.
33. B. Magill, C. Rohrs, and R. Stevenson. Output-queued switch emulation by fabrics with limited memory. *IEEE Journal on Selected Areas in Communications*, May:606–615, 2003.
34. A. Mekkittikul. *Scheduling Non-Uniform Traffic In High Speed Packet Switches And Routers*. PhD thesis, Stanford University, Nov 1998.
35. L. Mhamdi and M. Hamdi. Practical scheduling algorithms for high-performance packet switches. *ICC*, pages 1659–1663, May 2003.
36. L. Mhamdi and M. Hamdi. MCBF: A high-performance scheduling algorithm for internally buffered crossbar switches. *IEEE Communications Letters*, 7(9):451–453, Sep 2003.
37. K. Pun and M. Hamdi. Distro: A distributed static round-robin scheduling algorithm for bufferless clos-network switches. *GLOBECOM*, Nov 2002.
38. F.M. Chiussi and A. Francini. A distributed scheduling architecture for scalable packet switches. *IEEE Journal on Selected Areas in Communications*, 18(12):2665–2683, Dec 2000.
39. E. Oki, Z. Jing, R. Rojas-Cessa, and J. Chao. Concurrent round-robin dispatching scheme in a clos-network switch. *ICC*, 1:107–111, Jun 2001.
40. K. Pun and M. Hamdi. Static round-robin dispatching schemes for clos-network switches. *HPSR*, pages 329–333, May 2002.
41. C. Minkenberg, R. P. Luijten, F. Abel, W. Denzel, and M. Gusat. Current issues in packet switch design. *ACM SIGCOMM Comput. Commun. Review*, 33:119–124, 2003.
42. F. Masetti, P. Gavignet-Morin, D. Chiaroni, and G. Da Loura. Fiber delay lines optical buffer for ATM photonic switching applications. In *Proc. of INFOCOM*, pages 935–942, April 1993.
43. L. A. Buckman, K. S. Giboney, J. Straznicky, J. Simon, S. W. Corzine, X. J. Zhang, A. J. Schmit, and D. W. Dolfi. Parallel optical interconnects. In *Proc. Conf. Lasers and Electro-Optics (CLEO)*, pages 535–536, San Francisco, CA, May 2000.
44. S. Bregni, A. Pattavina, and G. Vegetti. Architectures and performance of AWG-based optical switching nodes for IP networks. *IEEE J. Select. Areas Commun.*, 21:1113–1121, September 2003.

45. H. J. Chao, K. L. Deng, and Z. Jing. Petastar: a petabit photonic packet switch. *IEEE J. Select. Areas Commun.*, 21:1096– 1112, September 2003.
46. L. Dittmann, C. Develder, D. Chiaroni, F. Neri, F. Callegati, W. Koerber, A. Stavdas, M. Renaud, A. Rafel, J. Sole-Pareta, Cerroni, N. Leligou, L. Dembeck, B. Mortensen, M. Pickavet, N. Le Sauze, M. Mahony, B. Berde, and G. Eilenberger. The European IST project DAVID: a viable approach toward optical packet switching. *IEEE J. Select. Areas Commun.*, 21:1026–1040, September 2003.
47. K. Kar, D. Stiliadis, T. V. Lakshman, and L. Tassiulas. Scheduling algorithms for optical packet fabrics. *IEEE J. Select. Areas Commun.*, 21:1143–1155, September 2003.
48. X. Li and M. Hamdi. On scheduling optical packet switches with reconfiguration delay. *IEEE J. Select. Areas Commun.*, 21:1156–1164, September 2003.
49. P. B. Chu, S. S. Lee, and S. Park. MEMS: the path to large optical crossconnects. *IEEE Commun. Mag.*, 40:80–87, 2002.
50. Tze-Wei Yeow, K. L. E. Law, and A. Goldenberg. MEMS optical switches. *IEEE Commun. Mag.*, pages 158–163, November 2001.
51. M. Hoffman, P. Kopka, and E. Voges. Thermooptical digital switch arrays in silica on silicon with defined zero voltage state. *J. Lightwave Technol.*, pages 395–400, March 1998.
52. J. E. Fouquet, S. Venkatesh, M. Troll, D. Chen, H. F. Wong, , and P. W. Barth. A compact, scalable cross-connect switch using total internal reflection due to thermally-generated bubbles. In *Proc. IEEE/LEOS Annu. Meeting*, pages 169–170, Orlando, FL, May 1988.
53. X. H. Ma and G. H. Kuo. Optical switching technology comparison: optical MEMS vs. other technologies. *IEEE Commun. Mag.*, 41:S16–S23, 2003.
54. P. D. Dobbelaere, K. Falta, S. Gloeckner, and S. Patra. Digital MEMS for optical switching. *IEEE Commun. Mag.*, 40:88–95, 2002.
55. L. Y. Lin, E. L. Goldstein, and R. W. Tkach. Free-space micromachined optical switches for optical networking. *IEEE J. Select. Topics Quantum Electron.*, 5:4–9, 1999.
56. T. Inukai. An efficient SS/TDMA time slot assignment algorithm. *IEEE Trans. Commun.*, 27:1449–1455, October 1979.
57. I. S. Gopal and C. K. Wong. Minimizing the number of switchings in an SS/TDMA system. *IEEE Trans. Commun.*, 33:497–501, June 1985.
58. B. Towles and W. J. Dally. Guaranteed scheduling for switches with configuration overhead. *IEEE/ACM Trans. Networking*, 11:835–847, October 2003.
59. R. Cole and J. Hopcroft. On edge coloring bipartite graphs. *SIAM J. on Comput.*, 11:540–546, August 1982.

2

Theoretical Performance of Input-queued Switches using Lyapunov Methodology

Andrea Bianco, Paolo Giaccone, Emilio Leonardi, Marco Mellia, and Fabio Neri

Dipartimento di Elettronica, Politecnico di Torino, C.so Duca degli Abruzzi 24, Torino, Italy
firstname.lastname@polito.it

The stochastic Lyapunov methodology is a powerful analytical tool used to assess the performance of queueing systems. We discuss here the main results obtained applying this methodology to the context of packet switch architectures with input queues, considering both an isolated switch and networks of switches. The methodology allows one to assess the limit performance in terms of throughput and delay; as a consequence, it is used to devise optimal scheduling algorithms. The theoretical results presented here can be applied to the practical design of high-speed packet switches.

2.1 Introduction

In recent years, much attention has been devoted by the research community to the design of input-queued (IQ) packet switching architectures and to the assessment of their performance.

An IQ switch architecture, depicted in Figure 2.1, is usually built around a non-blocking bufferless switching fabric and interconnecting input to output lines; examples of such fabric are crossbars, Clos networks, Benes networks, and Batcher–Banyan networks. At any time the switching fabric can be configured to provide a

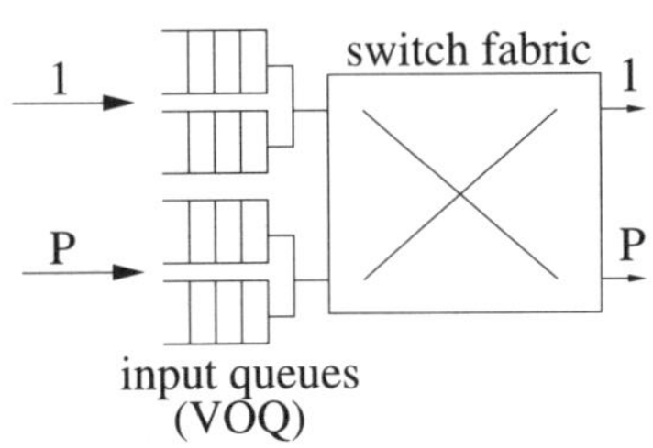

Figure 2.1. Basic architecture of a $P \times P$ IQ switch

set of parallel input/output connections, later called "matching", such that each input (output) is connected with at most one output (input). All switching fabric connections run at the speed of the external lines (normally assumed to be equal for all inputs and outputs); indeed, this allows the design of high-speed switching fabrics. Buffering and local processing is available for each external line in a line card that provides termination of the line and interfacing to the fabric. In the case of contention, packets are stored in buffers at the input line cards. IQ switches have become an attractive architectural solution for the design of large-size and high-capacity packet switches since seminal work [4, 23, 30, 31] showed that the negative effects on performance of Head-of-the-Line (HoL) blocking [16] can be reduced or completely eliminated by adopting per-destination queueing (also called Virtual Output Queueing - VOQ) at input cards.

A major issue in the design of IQ packet switches is that access to the switching fabric must be controlled by a scheduling algorithm, which operates on a (possibly partial) knowledge of the state of input queues. This means that control information must be exchanged among line cards, either through an additional data path, or through the switching fabric itself, and that significant processing power must be devoted to the scheduling algorithm, either at a centralized scheduler, or at line cards in a distributed manner.

We refer in this chapter to the case of fixed-size data units, called "cells", possibly obtained by segmenting variable-size packets (for example IP datagrams), and to synchronous switch operation, according to which input/output connections are changed synchronously at every cell time (called "slot") for all ports. Cell-based designs have been quite popular, as they permit one to reduce the complexity for both the hardware architecture and the scheduling algorithm.

The problem faced by scheduling algorithms with VOQ can be formalized as maximum size or maximum weight matching on a bipartite graph in which nodes represent input and output ports, and edges represent cells to be switched. Edges may be associated with weights related to the state of input queues. If P is the number of ports, then the total number of possible switching configurations (matchings) is $P!$, corresponding to the number of input/output permutations.

To achieve good scalability in terms of switch size and port data rate, it is essential to reduce the computational complexity of the scheduling algorithm. But a simpler algorithm may exhibit reduced performance. Hence, a possible solution is to introduce moderate speedup with respect to the data rate of input/output lines [9] in the switching fabric connections, as well as in the input and output memories. In this case, buffering is required at outputs as well as inputs, and the term "Combined Input/Output Queueing" (CIOQ) is used. Obviously, when the speedup is such that the internal switch bandwidth equals the sum of the data rates on input lines, input buffers become useless; in this case, the architecture becomes purely output-queued (OQ).

Along with the search for low complexity, highly scalable, well performing, switch architectures and scheduling algorithms, effort has recently been devoted to the identification and development of analytical methodologies to assess the performance achievable by IQ and CIOQ switch architectures. A complete set of general theoretical results could indeed provide an important framework to drive applied researchers towards better performing solutions. To this end, the stochastic Lyapunov function methodology played a fundamental role, since it permitted one to obtain most of the known theoretical results about the throughput of IQ and CIOQ switches. In addition, the methodology was applied in the wider scenario of a network of switches, providing new insights into their performance.

In this chapter we first briefly introduce the stochastic Lyapunov function methodology, showing how it can be successfully applied to determine the stability region of a system of queues; then, we summarize the main results about throughput performance of IQ and CIOQ architectures. We also show how the stochastic Lyapunov function methodology can be successfully applied to obtain bounds on the packet delay in IQ and CIOQ switches. Finally, we discuss the main results obtained for a network of switches.

2.2 Theoretical Framework

We introduce the theoretical framework to describe all the results regarding both switches in isolation and networks of switches, starting from the notation of a generic queueing system.

2.2.1 Description of the Queueing System

Consider a system of J discrete-time queues (of infinite capacity) represented by[1] vector Q, whose jth component, $0 \leq j < J$, is a descriptor associated with the jth queue in the system. The system of queues handles N classes of customers. Each customer arrives at the network from outside, receives service at a number of queues, and leaves the network. Customers change class every time they move through the network. We suppose that each class k of customers, $0 \leq k < N$, univocally identifies a queue in the system at which all class k customers are queued, *i.e.* all customers of class k are queued at the same queue; then $N \geq J$. Let $L(k) = j$ be the system location function that associates each class k of customers with the queue j at which class k customers are queued. $L^{-1}(j)$ is the counter-image of j through function $L(k)$. In general $L^{-1}(j)$ returns a set of customer classes. When $N = J$, each customer class is in one-to-one correspondence with a queue.

Let $X_n = (x_n^{(0)}, x_n^{(1)}, \dots, x_n^{(N-1)})$ be the vector whose kth component $x_n^{(k)}$, $0 \leq k < N$, represents the number of customers of class k in the system at time n. We say that the set of customers of the same class forms a virtual queue in the system of queues; thus we indicate the set of customers of class k with the term "virtual queue k". We suppose that the service times required by customers of all classes are deterministic and equal to one timeslot. We consider only nonpre-emptive atomic service policies, *i.e.* service policies that serve customers in an atomic fashion, never interrupting the service of the customer that is currently in service.

[1] All vectors are defined as row vectors.

The evolution of the number of queued customers is described by $x^{(k)}_{n+1} = x^{(k)}_n + e^{(k)}_n - d^{(k)}_n$, where $e^{(k)}_n$ represents the number of class k customers entering virtual queue k (and thus physical queue $L(k)$) in time interval $(n, n+1]$, and $d^{(k)}_n$ represents the number of customers departing from virtual queue k in time interval $(n, n+1]$. $E_n = (e^{(0)}_n, e^{(1)}_n, \ldots, e^{(N-1)}_n)$ is the vector of entrances in the virtual queues, and $D_n = (d^{(0)}_n, d^{(1)}_n, \ldots, d^{(N-1)}_n)$ is the vector of departures from the virtual queues. With this notation, the system evolution equation can be written as

$$X_{n+1} = X_n + E_n - D_n \tag{2.1}$$

The entrance vector is the sum of two terms: vector $A_n = (a^{(0)}_n, a^{(1)}_n, \ldots, a^{(N-1)}_n)$ representing the customers arriving at the system from outside, and vector $T_n = (t^{(0}_n, t^{(1)}_n, \ldots, t^{(N-1)}_n)$ of recirculating customers; $t^{(k)}_n$ is the number of customers departing from some virtual queue and entering virtual queue k in time interval $(n, n+1]$. Note that when customers do not traverse more than one queue (as is typically the case for a switch in isolation), vector T_n is null for all n, and $A_n = E_n$.

For simplicity of notation, we assume static routing (the extension to the case of dynamic routing is presented in [3]). The $N \times N$ matrix $R = [r^{(k,l)}]$ is the *routing matrix,* whose binary element $r^{(k,l)} = 1$ iff a customer served at virtual queue k is moved to virtual queue l. We assume that the system of queues forms an *open network, i.e.*[2] $I + R + R^2 + R^3 + \ldots = (I - R)^{-1}$ exists and is finite, *i.e.* $I - R$ is invertible. Note that $T_n = D_n R$. The law of evolution of virtual queues can thus be rewritten as

$$X_{n+1} = X_n + A_n - D_n(I - R) \tag{2.2}$$

Let us consider the external arrival process $A_n = (a^{(0)}_n, a^{(1)}_n, \ldots, a^{(N-1)}_n)$; we suppose that arrival processes are stationary, *i.e.* $E[A_n] = \Lambda = (\lambda^{(0)}, \lambda^{(1)}, \ldots, \lambda^{(N-1)})$ does not depend on the time interval $[n, n+1)$.

The average workload $E[W_n]$ provided at each virtual queue by customers that in time interval $[n, n+1)$ entered the system of queues is given by $E[W_n] = \Lambda(I - R)^{-1}$.

Before proceeding, we recall some norm functions that will be helpful in the sequel.[3]

Definition 1 *Given a vector* $Z \in I\!R^N$, $Z = (z^{(k)}, 0 \le k < N)$, *norm* $\|Z\|_p$ *is defined as*

$$\|Z\|_p = \left(\sum_{k=0}^{N-1} |z^{(k)}|^p \right)^{1/p}$$

---

[2] $E[X]$ denotes the expectation of random quantity X. I denotes the identity matrix, whose elements are equal to 1 on the diagonal, and null everywhere else.

[3] Here, $I\!N$ denotes the set of non-negative integers, $I\!R$ denotes the set of real numbers, and $I\!R^+$ denotes the set of non-negative real numbers.

Definition 2 *Given a location function $L(k) = j$, from $0 \leq k < N$ to $0 \leq j < J$, with $J \leq N$, norm $||Z||_{\mathrm{maxL}}$ (the name refers to maximum queue length) is defined as*

$$||Z||_{\mathrm{maxL}} = \max_{j=0,\ldots,J-1} \left\{ \sum_{k \in L^{-1}(j)} |z^{(k)}| \right\} \tag{2.3}$$

2.2.2 Stability Definitions for a Queueing System

Several definitions of stability for a network of queues can be found in the technical literature. We recall here some of them.

Definition 3 *Under a stationary exogenous arrival process $\{A_n\}$ satisfying the strong law of large numbers, i.e.:*

$$\lim_{n \to \infty} \frac{\sum_{i=0}^{n-1} A_i}{n} = \Lambda \qquad \textit{with probability 1}$$

A system of queues is rate stable if

$$\lim_{n \to \infty} \frac{X_n}{n} = \lim_{n \to \infty} \frac{1}{n} \sum_{i=0}^{n-1} (E_i - D_i) = 0 \quad \textit{with probability 1}$$

where X_n is the vector of queue sizes at time n.

Definition 4 *Under a stationary exogenous arrival process $\{A_n\}$, a system of queues is* weakly stable *if, for every $\epsilon > 0$, there exists $B > 0$ such that*

$$\lim_{n \to \infty} P\{||X_n|| > B\} < \epsilon$$

where $P\{E\}$ denotes the probability of event E.

Definition 5 *Under a stationary exogenous arrival process $\{A_n\}$, a system of queues is* strongly stable *if*

$$\lim_{n \to \infty} \sup E[||X_n||] < \infty$$

Any norm can be used in the two definitions above.

Note that strong stability implies weak stability, and that weak stability implies rate stability. Indeed, the rate stability property allows queue sizes to grow indefinitely with sub-linear rate, while the weak stability property entails that the servers in the system of queues process the whole offered load, but the delay experienced by customers can be unbounded. Strong stability implies, in addition, the boundness of average queue sizes and customer delays.

A necessary condition for the system of queues to achieve stability is that the average workload provided at each queue by customers entering the system of queues in time interval $[n, n+1)$ does not reach 1. This condition, that we call the *no-overload condition,* is also a sufficient condition for stability in any BCMP type network of queues [7]. This condition can be formalized as

$$||\Lambda(I-R)^{-1}||_{\text{maxL}} < 1 \tag{2.4}$$

In general, as shown in [6, 10, 11], this condition does not guarantee the stability of a generic network of queues. Although condition (2.4) can be extended to

$$||\Lambda(I-R)^{-1}||_{\text{maxL}} \leq 1$$

when rate stability is considered, we will normally refer to the stricter formulation (2.4) here.

2.2.3 Lyapunov Methodology

The systems under study can be modeled by discrete-time queues and they can be described with Discrete-Time Markov Chain (DTMC) models. Hence, we assume that the process describing the evolution of the system of queues is an irreducible DTMC, whose state vector at time n is $Y_n = (X_n, K_n)$, $Y_n \in \mathbb{N}^M$, $X_n \in \mathbb{N}^N$, $K_n \in \mathbb{N}^{N'}$, and $M = N + N'$. Y_n is the combination of vector X_n and a vector K_n of N' integer parameters. Let H be the state space of the DTMC, obtained as a subset of the Cartesian product of the state space H_X of X_n and the state space H_K of K_n.

From Definition 4, we can immediately see that if all states Y_n are positive recurrent, the system of queues is weakly stable; however, the converse is generally not true, since queue sizes can remain finite even if the states of the DTMC are not positive recurrent due to instability in the sequence of parameter $\{K_n\}$.

The following general criterion for the (weak) stability of systems that can be described with a DTMC is useful in the design of scheduling algorithms. This theorem is a straightforward extension of Foster's Criterion; see [13, 19, 33].

Theorem 1. *Given a system of queues whose evolution is described by a DTMC with state vector $Y_n \in \mathbb{N}^M$, if a lower bounded function $V(Y_n)$, called Lyapunov function, $V : \mathbb{N}^M \to \mathbb{R}$ can be found such that*[4] $E[V(Y_{n+1}) \mid Y_n] < \infty$, $\forall Y_n$, *and there exist $\epsilon \in \mathbb{R}^+$ and $B \in \mathbb{R}^+$ such that $\forall ||Y_n|| > B$*

$$E[V(Y_{n+1}) - V(Y_n) \mid Y_n] < -\epsilon \tag{2.5}$$

then all states of the DTMC are positive recurrent and the system of queues is weakly stable.

Note that an explicit dependence of the Lyapunov function on the time index n is allowed, so that it is possible to write explicitly $V(Y_n) = V(Y_n, n)$.

[4] We use the elementary notation for the conditional expectation, *i.e.* $E[X \mid Y \in A] = E[X, Y]/P\{A\}$, where A is an event set.

If the state space H of the DTMC is a subset of the Cartesian product of the denumerable state space H_X and a *finite* state space H_K, the stability criterion can be slightly modified, since the stability of the system can be inferred only from the queue size state vector X_n.

Corollary 1. *Given a system of queues whose evolution is described by a DTMC with state vector $Y_n \in I\!N^M$, and whose state space H is a subset of the Cartesian product of a denumerable state space H_X and a finite state space H_K, then, if a lower bounded function $V(X_n)$, called the Lyapunov function, $V : I\!N^N \to I\!R$ can be found such that $E[V(X_{n+1}) \mid Y_n] < \infty \quad \forall Y_n$ and there exist $\epsilon \in I\!R^+$ and $B \in I\!R^+$ such that $\forall Y_n : ||X_n|| > B$*

$$E[V(X_{n+1}) \quad V(X_n) \mid Y_n] < \quad \epsilon \tag{2.6}$$

then all states of the DTMC are positive recurrent.

In this case, the system of discrete-time queues is weakly stable iff all states of the DTMC are positive recurrent.

In the following, we restrict our analysis to systems of queues for which Corollary 1 applies. The following criterion for *strong* stability extends the previous result:

Theorem 2. *Given a system of queues whose evolution is described by a DTMC with state vector $Y_n \in I\!N^M$, and whose state space H is a subset of the Cartesian product of a denumerable state space H_X and a finite state space H_K, then, if a lower bounded function $V(X_n)$, called the Lyapunov function, $V : I\!N^N \to I\!R$ can be found such that $E[V(X_{n+1}) \mid Y_n] < \infty \quad \forall Y_n$ and there exist $\epsilon \in I\!R^+$ and $B \in I\!R^+$ such that $\forall Y_n : ||X_n|| > B$*

$$E[V(X_{n+1}) - V(X_n) \mid Y_n] < -\epsilon ||X_n|| \tag{2.7}$$

then the system of queues is strongly stable.

We report here the proof of the theorem, first derived in [21]. This proof is useful for readers interested in a practical application of the methodology; some intermediate mathematical steps will also be referred to later.

Proof. Since the assumptions of Theorem 1 are satisfied, every state of the DTMC is positive recurrent and the DTMC is weakly stable. In addition, to prove that the system is strongly stable, we shall show that $\lim_{n\to\infty} \sup E[||X_n||] < \infty$.

Let H_B be the set of values taken by Y_n for which $||X_n|| \leq B$ (where (2.7) does not apply). It is easy to prove that H_B is a compact set. Outside this compact set, Equation (2.7) holds, i.e.

$$E[V(X_{n+1}) - V(X_n) \mid Y_n] < -\epsilon ||X_n||$$

Considering all Y_n that do not belong to H_B, we obtain

$$E[V(X_{n+1}) - V(X_n) \mid Y_n \notin H_B] < -\epsilon E[||X_n|| \mid Y_n \notin H_B]$$

Instead, for $Y_n \in H_B$, with H_B a compact set,

$$E[V(X_{n+1}) \mid Y_n \in H_B] \leq M < \infty$$

where M is the maximum value taken by $E[V(X_{n+1}) \mid Y_n]$ for Y_n in H_B.

By combining the two previous expressions, we obtain

$$\begin{aligned} & E[V(X_{n+1})] < \\ & MP\{Y_n \in H_B\} + P\{Y_n \notin H_B\} \{E[V(X_n) \mid Y_n \notin H_B] - \epsilon E[||X_n|| \mid Y_n \notin H_B]\} < \\ & \qquad M + E[V(X_n)] - \epsilon E[||X_n||] + M_0 \end{aligned}$$

M_0 is a constant such that

$$M_0 \geq \{-E[V(X_n) \mid Y_n \in H_B] + \epsilon E[||X_n|| \mid Y_n \in H_B]\} P\{Y_n \in H_B\}$$

Note that M_0 is finite, H_B being a compact set. By summing over all n from 0 to $N_0 - 1$, we obtain

$$E[V(X_{N_0})] < N_0 M + E[V(X_0)] - \epsilon \sum_{n=0}^{N_0-1} E[||X_n||] + N_0 M_0$$

Thus, for any N_0, we can write

$$\frac{\epsilon}{N_0} \sum_{n=0}^{N_0-1} E[||X_n||] < M + \frac{1}{N_0} E[V(X_0)] - \frac{1}{N_0} E[V(X_{N_0})] + M_0$$

$E[V(X_{N_0})]$ is lower bounded by definition; assume $E[V(X_{N_0})] > K_0$. Hence

$$\frac{\epsilon}{N_0} \sum_{n=0}^{N_0-1} E[||X_n||] < M + \frac{1}{N_0} E[V(X_0)] - \frac{K_0}{N_0} + M_0$$

For $N_0 \to \infty$, being $E[V(X_0)]$ and K_0 finite, we can write

$$\frac{\epsilon}{N_0} \sum_{n=0}^{N_0-1} E[||X_n||] < M + M_0 \tag{2.8}$$

Hence $\lim_{N_0 \to \infty} \frac{1}{N_0} \sum_{n=0}^{N_0-1} E[||X_n||]$ is bounded. Since the DTMC Y_n has positive recurrent states, there exists $\lim_{n \to \infty} E[||X_n||]$. Furthermore, if the sequence $E[||X_n||]$ is convergent, the sequence $\frac{1}{n} \sum_{i=0}^{n-1} E[||X_i||]$ converges to the same limit (the Cesaro sum):

$$\lim_{n \to \infty} E[||X_n||] = \lim_{n \to \infty} \frac{1}{n} \sum_{i=0}^{n-1} E[||X_i||]$$

But the right-hand side is seen to be bounded; hence, $\lim_{n \to \infty} E[||X_n||] < \infty$

A class of Lyapunov functions is of particular interest:

Corollary 2. *Given a system of queues as in Theorem 2, then, if there exists a symmetric copositive* [5] *matrix* $Z \in I\!R^{N \times N}$, *and two positive real numbers* $\epsilon \in I\!R^+$ *and* $B \in I\!R^+$, *such that, given the function* $V(X_n) = X_n Z X_n^T$, $\forall Y_n : ||X_n|| > B$,

$$E[V(X_{n+1}) - V(X_n) \mid Y_n] < -\epsilon ||X_n|| \tag{2.9}$$

then the system of queues is strongly stable. In addition, all the polynomial moments of the queue size distribution are finite.

This is a re-phrasing of the results presented in [18, Section IV]. In particular, the identity matrix I is a symmetric positive semidefinite matrix, hence a copositive matrix; thus, it is possible to state the following:

Corollary 3. *Given a system of queues as in Theorem 2, if there exist* $\epsilon \in I\!R^+$, $B \in I\!R^+$ *such that* $\forall Y_n : ||X_n|| > B$

$$E[X_{n+1} X_{n+1}^T - X_n X_n^T \mid Y_n] < -\epsilon ||X_n|| \tag{2.10}$$

then the system of queues is strongly stable, and all the polynomial moments of the queue size distribution are finite.

2.2.4 Lyapunov Methodology to Bound Queue Sizes and Delays

The Lyapunov methodology is also useful to evaluate some bounds on average queue size and average delay of a single IQ switch, as described in [22, 29].

The key observation is that the proof of Theorem 2 provides a first bound on the limit behavior of $E[||X_n||]$. Indeed, from (2.8):

$$\lim_{n \to \infty} E[||X_n||] \leq \frac{1}{\epsilon}(M + M_0) \tag{2.11}$$

where M is the maximum taken by $E[V(X_{n+1}) \mid Y_n]$ for $Y_n \in H_B$, where $H_B = \{Y_n, ||X_n|| \leq B\}$, and M_0 is a constant such that

$$M_0 \leq \{-E[V(X_n) \mid Y_n \in H_B] + \epsilon E[||X_n|| \mid Y_n \in H_B]\} P\{Y_n \in H_B\}$$

Unfortunately, this bound is often very loose; thus, tighter bounds can be obtained by selecting special classes of Lyapunov functions.

Considering a system of queues satisfying the assumptions of Theorem 2, since the DTMC describing the evolution of the queues is positive recurrent, if we assume aperiodicity, the DTMC is ergodic. Moreover, since the system is strongly stable:

$$\lim_{n \to \infty} E[X_{n+1}] = \lim_{n \to \infty} E[X_n] < \infty$$

In addition, if the Lyapunov function $V(X_n)$ is a quadratic form, *i.e.* $V(X_n) = X_n Z X_n^T$, since all the polynomial moments of X_n are finite, it follows that:

---

[5] An $N \times N$ matrix Q is copositive if $XQX^T \geq 0 \quad \forall X \in I\!R^{+N}$.

$$\lim_{n\to\infty} E[V(X_{n+1}) - V(X_n)] = 0 \tag{2.12}$$

An extension to Theorem 2 can easily be obtained by replacing in (2.7) $-\epsilon||X_n||$ with $-\epsilon f(||X_n||)$, where $f(\cdot)$ is a continuous function defined on $\mathbb{R}^+$. In this case with steps similar to those in the proof of Theorem 2, it is possible to prove that $\lim_{n\to\infty} E[f(||X_n||)] < \infty$.

It is now possible to state the following theorem (from [22]) that provides a stronger and more general bound than (2.11).

Theorem 3. *Given a system of queues whose evolution is described by a DTMC with state vector $Y_n \in \mathbb{N}^M$, whose state space H is a subset of the Cartesian product of a denumerable state space H_X and a finite state space H_K, and for which all the polynomial moments of the queue sizes distribution are finite, if a lower bounded polynomial function $V(X_n)$, $V : \mathbb{N}^N \to \mathbb{R}$, can be found, such that*

$$E[V(X_{n+1}) \mid Y_n] < \infty \qquad \forall Y_n$$

and there exist two positive real numbers $\epsilon \in \mathbb{R}^+$ and $B \in \mathbb{R}^+$, such that

$$E[V(X_{n+1}) - V(X_n) \mid Y_n] \le -\epsilon f(||X_n||) \qquad \forall Y_n : ||X_n|| > B \tag{2.13}$$

with $f(x)$ a continuous function in $\mathbb{R}^+$, then

$$\lim_{n\to\infty} E[f(||X_n||)] \le$$
$$\lim_{n\to\infty} E\left[f(||X_n||) + \frac{V(X_{n+1}) - V(X_n)}{\epsilon} \mid Y_n \in H_B\right] P\{Y_n \in H_B\} \tag{2.14}$$

2.2.5 Application to a Single Queue

Consider a simple discrete-time (or time-slotted) $Geo^{(b)}/D/1$ queue with infinite buffer, where customers arrive in batches of size b at geometrically spaced time intervals, and require a deterministic service time (equal to one time slot). Such a queue can provide a simplified model for either a multiplexer of cell flows, or an output interface of an OQ cell switch, and can serve as an illustratory example of the methodology that we later apply to more complex queueing models of IQ and CIOQ cell switches.

Let x_n be the number of customers in the queue at time slot n; let a_n be the number of arrivals, and d_n the number of departures, during time slot n. Let $\lambda = E[a_n]$ be the average arrival rate. Observe that both $E[a_n^2]$ and λ are finite.

The equation describing the evolution of this system over time is

$$x_{n+1} = x_n + a_n - d_n \tag{2.15}$$

where

$$d_n = \begin{cases} 1 & \text{if} \quad x_n > 0 \\ 0 & \text{otherwise} \end{cases} \tag{2.16}$$

This simple queuing model corresponds to an irreducible discrete-time Markov chain (DTMC).

Now consider the following linear Lyapunov function: $V(x_n) = x_n$. If we fix $x_n > 0$, then we can write (2.5) as follows:

$$E[V(x_{n+1}) - V(x_n) \mid x_n > 0] = E[a_n - d_n] = \lambda - 1$$

Hence, thanks to Theorem 1, the queue is weakly stable for $\lambda < 1$.

Now consider the following quadratic Lyapunov function: $V(x_n) = x_n^2$. If we let $x_n \to \infty$, then we can write (2.7) as follows:

$$E[V(x_{n+1}) - V(x_n) \mid x_n] = E[2(a_n - d_n)x_n + (a_n - d_n)^2 \mid x_n] = 2(\lambda - 1)x_n + E[(a_n - 1)^2]$$

Now,

$$\lim_{x_n \to \infty} \frac{E[V(x_{n+1}) - V(x_n) \mid x_n]}{x_n} < -2(1 - \lambda) \tag{2.17}$$

and Theorem 2 proves the strong stability of the queue in the case $\lambda < 1$.

After evaluating (2.17), we can apply Theorem 3, using $f(||X_n||) = X_n$, to bound the average queue size $E[x]$. Letting $x_n \to \infty$:

$$\begin{aligned}
\lim_{x_n \to \infty} E[x_n] &\leq \lim_{x_n \to \infty} E\left[x_n + \frac{-2(1-\lambda)x_n + (a_n - d_n)^2}{2(1-\lambda)} \middle| x_n\right] \\
&= \lim_{x_n \to \infty} E\left[\frac{(a_n - d_n)^2}{2(1-\lambda)}\right] = E\left[\frac{a_n^2 - 2a_n d_n + d_n^2}{2(1-\lambda)}\right] \\
&= \frac{E[a_n^2] - 2\lambda^2 + \lambda}{2(1-\lambda)}
\end{aligned}$$

since $E[d_n] = \lambda$ because of ergodicity, and $E[d_n] = E[d_n^2]$ because d_n is a binary variable.

Note that the result obtained is the equivalent of the Pollaczek–Khinchin formula [17] for our discrete-time $Geo^{(b)}/D/1$ queue.

From queue size bounds, the derivation of bounds on the average cell delay is easy, thanks to Little's result.

The procedure for the derivation of bounds on the queue size variance is identical, but requires the use of a different function: $V(x_n) = x_n^3$. We omit the details.

2.2.6 Final Remarks

We notice that the stochastic Lyapunov methodology is a rather simple and powerful tool, which has been successfully applied to prove either the weak stability or strong stability of several complex systems of queues, such as IQ switches.

The application of stochastic Lyapunov function methodology imposes, however, some rather strong assumptions on the exogenous $\{A_n\}$ arrival process. Even if some extensions to more general cases are possible, usually the sequence of variables $\{X_n\}$, is required to be a DTMC; and thus, the exogenous arrival process at the system, $\{A_n\}$, needs to form an i.i.d. random variable sequence.

More advanced analytical tools, such as *fluid models*, can be applied to prove rate stability under relaxed assumptions on the exogenous arrival process $\{A_n\}$ (fluid models only require $\{A_n\}$ to be a stationary arrival process satisfying the strong law of large numbers).

2.3 Performance of a Single Switch

We consider IQ or CIOQ cell-based switches with P input ports and P output ports, all running at the same cell rate (and we call them $P \times P$ IQ or CIOQ switches). The switching fabric is assumed to be non-blocking and memoryless, *i.e.* cells are only stored at switch inputs and outputs.

At each input, cells are stored according to a Virtual Output Queueing (VOQ) policy: one separate queue is maintained at each input for each input–output couple. We do not model possible output queues since they never become unstable under admissible traffic patterns. The total number of input queues in each switch is $N = P^2$, which is also equal to the number of customer classes in the general queueing model: $J = N$.

The switch in isolation can be modeled as a system comprising N virtual queues. Let $q^{(k)}$, $k = Pi + j$ be the virtual queue at input i storing cells directed to output j, with $i, j = 0, 1, 2, \ldots, P-1$.

We define three functions referring to VOQ $q^{(k)}$:

- $I(k)$: returns the index of the input card in which the VOQ is located;
- $O(k)$: returns the index of the output card to which VOQ cells are directed.

We consider a synchronous operation, in which the switch configuration can be changed only at slot boundaries. We call *internal time slot* the time necessary to transmit a cell at an input port (or to receive it at an output port). We call *external time slot* the duration of a cell on input and output lines. The difference between external and internal time slots is due to switch speedup, and to possibly different cell formats (*e.g.* due to additional internal header fields).

At each internal time slot, the switch scheduler selects cells to be transferred from input queues to output queues. The set of cells to be transferred during an internal time slot must satisfy two constraints: (i) at most one cell can be extracted from the VOQ structure at each input, and (ii) at most one cell can be transferred towards each output, thus resulting in correlation among server activities at different queues.

In the following, we will discuss the stability properties for IQ switches and CIOQ switches with speedup 2; in addition, we will show some delay bounds for IQ switches. Before proceeding, we introduce some other mathematical notation.

We adapt the definition of $||Z||_{\mathrm{maxL}}$ to the case of the single switch.

Definition 6 *Given a vector* $Z \in \mathbb{R}^N$, $Z = (z^{(k)},\ k = Pi+j,\ i,j = 0,1,\ldots,P-1)$, *the norm* $||Z||_{IO}$ *is defined as:*

$$||Z||_{IO} = \max_{j=0,\ldots,P-1} \left\{ \sum_{k \in I^{-1}(j)} |z^{(k)}|,\ \sum_{k \in O^{-1}(j)} |z^{(k)}| \right\}$$

The constraint on the set of cells transferred through the switch can be formalized in the following manner.

Definition 7 *At each time slot, the scheduler of an IQ switch selects for transfer from queues* $Q = (q^{(k)})$ *a set of cells denoted by vector* $D \in \mathbb{N}^N$, $D = (d^{(k)} \in \{0,1\},\ k = Pi + j,\ i,j = 0,1,\ldots,P-1)$ *so that* $||D||_{IO} \leq 1$. *Set* D *is said to be a set of non-contending cells, or a switching vector.*

In order not to overload any input and output switch port, the total average arrival rates in cells/(external slot) must be less than 1 for all input and output ports; in this case we say that the traffic pattern is *admissible.*

Definition 8 *The traffic pattern loading an (isolated) IQ switch is admissible if and only if* $||E[E_n]||_{IO} = ||\Lambda||_{IO} < 1$.

Note that any admissible traffic pattern can be transferred without losses in an OQ switch architecture with infinite queues.

A traffic is said to be uniform if $\Lambda^{(i,j)} = \rho/P$, for $0 \leq i,j < P$, and $0 < \rho < 1$.

Finally we say that a system of queue achieves 100 % throughput when it is strongly stable under any admissible i.i.d arrival process.

2.3.1 Stability Region of Pure Input-queued Switches

We say that an IQ switch adopts a Maximum Weight Matching (MWM) scheduling policy if the selection of the switching vector in each slot is implemented according to the following rule:

$$D_n = \arg \max_{D_i \in \mathcal{D}_X} W_n D_i^T \tag{2.18}$$

where W_n is a vector of weights associated to VOQs, and $\mathcal{D}_X$ denotes the set of all possible $P!$ switching vectors at time n. Note that, for any $D \in \mathcal{D}_X$, $W_n D^T$ represents the weight of matching D.

When the policy maximizes the number of cells to transfer (corresponding to the binary case, *i.e.* $w_n^{(Pi+j)} = 1$ if the corresponding VOQ is not empty, 0 otherwise), the policy computes a maximum *size* matching.

Note also that a matching is said to be *maximal* when no other edge can be added, without violating the switching constraints $||D||_{IO} \leq 1$; in general, a maximal matching may not be maximum.

Pure IQ switches (*i.e.* switches with no speedup) implementing a MWM scheduling algorithm were proved to achieve the same performance in terms of throughput of OQ switches under a wide class of traffic patterns. This fundamental result was first obtained in [25, 26] under i.i.d. arrival processes, applying the stochastic Lyapunov function methodology, and then extended in [12] for more general arrival processes applying fluid models. To obtain maximum throughput, VOQ weights must be proportional to the queue size (*e.g.* $W_n = X_n$), or to the age of the head-of-the-line cell (Oldest Cell First policy) in the corresponding VOQ [26], or, finally, to the sum of all cells stored in the corresponding input and output ports (Longest Port First policy) [24].

In more recent years the previous results have been generalized in two main directions:

- works [3, 28] provide more general characterization of scheduling algorithms that guarantee 100% throughput in pure IQ architectures;
- works [15, 28, 34] exploit the system memory (*i.e.* the fact that X_n and X_{n+1} are strongly correlated) to simplify the scheduling algorithms while guaranteeing 100% throughput for IQ architectures.

In the following we report two results; the first, taken from [3], generalizes the result in [26] on MWM optimality; the second, taken from [34], proposes a simple algorithm that exploits system memory.

Definition 9 *Let $F(X)$ be a regular function[6] $F \in C^1[\mathbb{R}^{+N} \rightarrow \mathbb{R}^{+N}]$. An IQ switch adopts a* $F(X)$-max-scalar *scheduling policy if the selection of the switching vector in each slot is implemented according to the following rule:*

$$D_n = \arg \max_{D_i \in \mathcal{D}_{X_n}} F(X_n) D_i^T \tag{2.19}$$

where X_n is the vector of queue sizes, and $\mathcal{D}_{X_n}$ denotes the set of all possible switching vectors at time n.

Note that when $F(X) = X$, the above policy corresponds to the usual MWM. The following is the main stability result, which is an extension of [26].

Theorem 4. *Let $F(X)$ be a regular function $F \in C^1[\mathbb{R}^{+N} \rightarrow \mathbb{R}^{+N}]$ such that:*

1. *$F(X)$ defines a conservative field, i.e.*

$$\oint_{\Gamma} F(X) d\Gamma(X)^T = 0 \tag{2.20}$$

for each regular closed line Γ in $\mathbb{R}^{+N}$

2. *$F(X)$ grows to infinity when X grows to infinity; formally, there exists a finite $s > 0$ such that:*

$$\liminf_{||X|| \rightarrow \infty} \frac{||F(X)||}{||X||} \geq s \tag{2.21}$$

3. *all null elements of X remain null:*

$$U[X] F(X) = F(X) \tag{2.22}$$

[6] C^n denotes the set of continuous functions with continuous ith derivative, $1 \leq i \leq n$.

Then an IQ switch adopting the $F(X)$-max-scalar policy is strongly stable under any admissible i.i.d. traffic pattern.

Proof. Let us define the function $\mathcal{L}(X)$:

$$\mathcal{L}(X) = \int_{\Gamma_X} F(Y) d\Gamma_X(Y)^T \tag{2.23}$$

$$\mathcal{L}(0) = 0 \tag{2.24}$$

where Γ_X is an open regular line with endpoints 0 and X.

By definition $\mathcal{L}(X) \in C^2[I\!R^{+N} \to I\!R]$. It is easy to verify that, for each $X \in I\!R^{+N}$, $\mathcal{L}(X) \geq 0$. To see this, it is sufficient to consider a straight line Γ_X parallel to vector X. Since $X \in I\!R^{+N}$, both $F(Y)$ and $d\Gamma_X(Y)$ in (2.23) belong to $I\!R^{+N}$ for all Y, so that also $\mathcal{L}(X) \in I\!R^{+N}$.

Let us consider $\mathcal{L}(X)$ as our Lyapunov function. Since the maximum number of cells arriving in a slot at the switch is bounded, then $||X_{n+1}||_2$ is bounded for any finite X_n, and from the regularity of $\mathcal{L}(X)$ follows that

$$E[\mathcal{L}(X_{n+1}) \mid X_n] < \infty$$

Finally, for $||X_n||_2 \to \infty$, by writing a Taylor series for $\mathcal{L}(X_n + A_n - D_n) = \mathcal{L}(X_n) + \nabla\mathcal{L}(X_n)(A_n - D_n)^T + \ldots$, we obtain:

$$\begin{aligned} \frac{E[\mathcal{L}(X_{n+1}) \mid X_n] - \mathcal{L}(X_n)}{||X_n||_2} &= O\left(\frac{\nabla\mathcal{L}(X_n)(E[A] - D_n)^T}{||X_n||_2}\right) \\ &= O\left(\frac{F(X_n)(E[A] - D_n)^T}{||X_n||_2}\right) \end{aligned} \tag{2.25}$$

We must now show that (2.25) is smaller than a negative finite constant. By the Birkhoff-von-Neumann theorem [8], every vector Y in $I\!R^{+N}$ such that $||Y||_{IO} \leq 1$ belongs to the convex hull of the switching vectors. Since the arrival process is admissible, hence it is internal to the convex hull generated by departure vectors ($||A||_{IO} < 1$), there exists an $\epsilon > 0$, and a vector $A' = E[A] + \epsilon D_n$, $A' \in I\!R^{+N}$, which is again internal to the convex hull ($||A'||_{IO} < 1$). We can write $E[A] = A' - \epsilon D_n$, and substitute in the right-hand side of (2.25), whose numerator becomes $[F(X_n)(A' - \epsilon D_n - D_n)^T]$. Now, by the linearity of functional $F(X_n)Y^T$ with respect to Y^T, and the definition of $F(X)$-max-scalar policy, it follows that, under the assumptions of the theorem, $F(X_n)A' \leq \max_{D^* \in \mathcal{D}_{X_n}} F(X_n)D^{*T} = U[X_n]F(X_n)D_n^T$, thus:

$$\frac{E[\mathcal{L}(X_{n+1})|X_n] - \mathcal{L}(X_n)}{||X_n||_2} \leq -\epsilon \frac{F(X_n)D_n^T}{||X_n||_2}$$

Then, for $||X_n||_2$ growing to infinity, using (2.21) and the fact that $||D_n||$ is always finite,

$$\frac{E[\mathcal{L}(X_{n+1})|X_n] - \mathcal{L}(X_n)}{||X_n||_2} < -\epsilon'$$

where ϵ' is a positive constant depending on N and $F(X)$.

Coming back to optimal throughput algorithms which exploit system memory, we present the landmark policy originally proposed in [34]. Notice that several better engineered policies exploiting system memory have been proposed later [15, 28]; however due to lack of space we refer to the original papers for this class of policies.

The intuition of the approach in [34] is that, if we assume that weights correspond to queue sizes, then correlation exists between the MWM computed in subsequent time slots; indeed, if D_n is the maximum weight matching computed at time n, it can easily be shown that

$$|X_n D_n^T - X_{n-1} D_{n-1}^T| \leq 2P$$

This correlation can be exploited by memorizing the matching used in the previous time slot; if the previous matching was optimal, now it can be considered a good "guess" of the current optimal matching. In addition, the approach exploits randomness to search, at any time, a possible MWM.

This policy, as originally defined in [34], can be formalized as follows. Let I_n be a random matching chosen among all $P!$ possible through, for example, a uniform distribution. Now use D_n as the matching, chosen between I_n and D_{n-1}, with the maximum weight:

$$D_n = \arg \max_{D \in \{I_n, D_{n-1}\}} X_n D^T$$

[33] proves that this policy guarantees 100% throughput.

2.3.2 Delay Bounds for Maximal Weight Matching

Stability proved through Lyapunov methodology is asymptotic, since negative drift is shown in the region $\|X_n\| \to \infty$. Unfortunately, this region has very limited practical relevance, unless some queue sizes or delay bounds are available.

Bounds on the average queue size (and cell delay, as a consequence of Little's law) in a switch implementing MWM with $W_n = X_n$ was obtained in [22] applying a particular polynomial Lyapunov functions $V(X_n)$. By applying Theorem 3, in [22] it is shown that

$$E[||X_n||_1] \leq \frac{||\Lambda||_1 - ||\Lambda||_2^2}{1 - ||\Lambda||_{IO}} P \tag{2.26}$$

In the case of uniform traffic, a bound on the average size of individual input queues can easily be obtained:

$$E[x^{(i)}] \leq \frac{\rho - \rho^2/P}{1 - \rho} \tag{2.27}$$

where $\rho = ||\Lambda||_{IO}$ represents the port load. Then by Little's theorem the average cell delay is obtained:

$$E[T] = \frac{E[x^{(i)}]}{E[a^{(i)}]} \leq \frac{P - \rho}{1 - \rho} \tag{2.28}$$

with $E[a^{(i)}] = \rho/P$.

Other stochastic methodologies, as in [5, 27], have been applied to provide delay bounds.

2.3.3 Stability Region of CIOQ with Speedup 2

The implementation of optimal schedulers in pure IQ switches can be rather problematic since their algorithmic complexity is rather large (the execution of MWM requires at least $O(P^3)$, see [32]).

However, simpler schedulers may lead to optimal performance if the switch is provided with a moderate speedup $S > 1$. This consideration has encouraged researchers to look for simple and efficient scheduling policies which can easily be implemented in high bandwidth routers. These policies usually compute a maximal size matching, as in the case of WFA [31], iSLIP [23], 2DRR [20], and many others recently proposed.

An important theoretical investigation of these policies is provided in [12, 21], where it has been proved that CIOQ switches with speedup 2 implementing any maximal Size Matching scheduling (mSM) algorithm achieve 100% throughput.

Consider any queue $q^{(k)},\ k = Pi + j$, in the VOQ structure, that stores cells at input i directed to output j. Recall that cells stored in $q^{(k)}$ compete for inclusion in the set of non-contending cells with cells stored in each queue $q^{(k')},\ k' = Pi + l$ with $l \neq j$, and $q^{(k'')},\ k'' = Ph + j$ with $h \neq i$.

If $q^{(k)}$ is non-empty, the mSM algorithm generates a set of non-contending cells that comprises at least one cell extracted from the *interfering set* $I^{(k)}$

$$I^{(k)} = \bigcup \{q^{(k')} \cup q^{(k'')}\} \tag{2.29}$$

(exactly one, if the cell is extracted from $q^{(k)}$; possibly two, if one cell is extracted from a $q^{(k')}$, $k' \neq k$, and one from a $q^{(k'')}$, $k'' \neq k$).

Because CIOQ switches have both input and output queues, in the sequel we will use X_n to indicate the state of the input VOQs, while $O_n \in \mathbb{N}^P$ indicates the state vector of output queues.

Theorem 5. *A CIOQ switch with speedup $S \geq 2$ adopting an mSM scheduler is strongly stable under any admissible traffic pattern.*

The stability of mSM scheduling algorithms with $S = 2$ was first proved under a weaker sense (rate stability) in [12] applying the fluid models methodology, and then strengthened in [21, 22] applying the stochastic Lyapunov function methodology. The proof we report is taken from [22].

Proof. Consider as Lyapunov function $V(X_n) = X_n Q X_n^T$, where $Q \in \mathbb{R}^N \times \mathbb{R}^N$ is such that

$$q_{ij} = \begin{cases} 1 & \text{if} \quad j = (i + lP) \bmod N, \ \ l = 0, \ldots, P-1 \\ 1 & \text{if} \quad P\lfloor i/P \rfloor \leq j < P\lfloor i/P \rfloor + P \\ 0 & \text{otherwise} \end{cases}$$

i.e. $X_n Q = \sum_{l \in I^{(k)}} x_n^{(l)}$ is the number of cells stored in interfering queues. Figure 2.2 reports a sketch of the Q matrix structure.

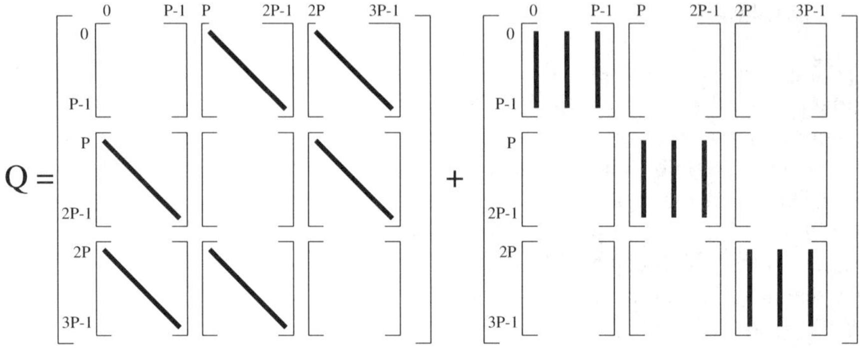

Figure 2.2. Schematic block for the Q operator used as Lyapunov function

We need to prove that inequality (2.7) holds to prove system stability. Thus, recalling that $X_{n+1} = X_n + A_n - \mathcal{D}_n$, we have

$$
\begin{aligned}
&E[V(X_{n+1}) - V(X_n) \mid X_n] \\
&= E[2X_n Q(A_n - \mathcal{D}_n)^T + (A_n - \mathcal{D}_n)Q(A_n - \mathcal{D}_n)^T \mid X_n]
\end{aligned}
$$

being Q a symmetric matrix. For $||X_n||_1 \to \infty$, it holds that

$$
\begin{aligned}
E[2X_n Q(A_n - \mathcal{D}_n)^T &+ (A_n - \mathcal{D}_n)Q(A_n - \mathcal{D}_n)^T \mid X_n] \\
&= 2E[X_n Q A_n^T \mid X_n] - 2E[X_n Q \mathcal{D}_n^T \mid X_n] + o(||X_n||_1)
\end{aligned}
$$

where $\lim_{||X_n||_1 \to \infty} \frac{o(||X_n||_1)}{||X_n||_1} = 0$. But $E[A_n]Q \leq ||\Lambda||_{\text{maxL}} \mathbb{I}$ and

$$
E[X_n Q A_n^T \mid X_n] = E[A_n] Q X_n^T \leq ||\Lambda||_{\text{maxL}} ||X_n||_1 \tag{2.30}
$$

At the same time, for the definition of the mSM algorithm, which selects $\mathcal{D}_n$ comprising at least one cell from interfering set $I^{(k)}$, and given the speedup $S \geq 2$,

$$
\begin{aligned}
(E[D_n]Q)^{(k)} X_n^{(k)} &\geq 2x_n^{(k)} - 1P\{x_n^{(k)} = 1 \wedge d_n^{(k)} = 1\} \\
&\geq 2x_n^{(k)} - P\{d_n^{(k)} = 1\} = 2x_n^{(k)} - E[d_n^{(k)}] = 2x_n^{(k)} - \lambda^{(k)}
\end{aligned}
$$

Thus

$$
E[X_n Q \mathcal{D}_n^T \mid X_n] \geq 2||X_n||_1 - ||\Lambda||_1 \tag{2.31}
$$

Thus, there exist $B > 0$ and $\epsilon > 0$ such that:

$$
\begin{aligned}
E[V(X_{n+1}) &- V(X_n) \mid X_n] \\
&= 2E[X_n Q A_n^T \mid X_n] - 2E[X_n Q \mathcal{D}_n^T \mid X_n] + o(||X_n||_1) \\
&\leq -\epsilon ||X_n||_1 \qquad \forall X_n : ||X_n||_1 > B
\end{aligned}
$$

Finally, we emphasize that several other scheduling policies which do not fall into the class of mSM have been shown to achieve optimal throughput in switches with speedup $S \geq 2$ [21]. As an example, any scheduling policy according to which the selected departure vector D_n is such that

$$X_n D_n^T > \frac{1}{2} \max_{D \in \mathcal{D}_X} X_n D^T$$

was proved to achieve optimal throughput in switches with speedup $S \geq 2$ [21].

2.3.4 Scheduling Variable-size Packets

All the results shown so far through the Lyapunov methodology assume that fixed-size cells are switched across the switching fabric. To apply these results in the context of an IP router, the hardware architecture must include some modules which, at the inputs, chop the variable-size packets into fixed-size cells and, at the outputs, reassemble the cells. Even if this architecture is common in practice, it requires additional complexity to handle the conversion between packets and cells; hence, other architectures have been designed to switch variable-size packets. Fortunately, some theoretical results have been produced in recent years on the achievable throughput of IQ switches handling variable-size packets [1, 14].

Variable-length packets are modeled in this context as trains of fixed-size cells which have to be transferred to output ports through synchronous fabrics in contiguous time slots. In paper [1], applying the stochastic Lyapunov function methodology, it has been proved that a pure IQ switch with no speedup, implementing a variant of the MWM algorithm allowing the transfer of cells originated by the same packet in contiguous time slots, still achieves 100% throughput.

More precisely, denote by $\mathcal{S}_n$ the set of VOQs from which the transfer of a packet occurs at time slot n.

Theorem 6. *An IQ switch with no speedup is strongly stable under any admissible traffic pattern if at each time slot the departure vector is selected according to a MWM scheduling algorithm*

$$D_n = \arg \max_{D_i \in \mathcal{D}_X} W_n D_i^T \tag{2.32}$$

in which VOQ weights are

$$w_n^{(k)} = \begin{cases} x_n^{(k)} & \text{if } k \notin \mathcal{S}_n \\ \infty & \text{if } k \in \mathcal{S}_n \end{cases}$$

This result was extended in [14] under a wider class of arrival processes by applying the fluid model methodology.

2.4 Networks of IQ Switches

Consider the case of a network interconnecting many IQ switches, each of them running a MWM scheduling algorithm, which guarantees to achieve 100% throughput when each switch is studied in isolation. It was shown in [6] that a specific network of IQ switches implementing a MWM scheduling policy can exhibit unstable behavior even when none of the switches are overloaded. This counterintuitive result opened new perspectives in the research on IQ and CIOQ switches, reducing the value of most of the results obtained for switches in isolation. In [6] the authors propose a policy named LIN that, if implemented in each switch of the network, leads to 100% throughput under any admissible traffic pattern when each traffic flow in the network is leaky-bucket compliant. The LIN policy, however, is based on pre-scheduling cell transmissions at each switch in the network, thus relying on exact knowledge of the traffic pattern at each switch, an approach not feasible in practice. In addition, the result proved in [6] cannot easily be extended to more general traffic patterns in which flows are not leaky-bucket compliant.

As an example of counterintuitive behavior, consider the network of eight IQ switches depicted in Figure. 2.3, in which continuous lines represent links between switches, and dashed lines represent information flows and their routing in the network. Note that each pair of adjacent IQ switches (all pairs are alike) is traversed by a locally originated flow, a locally terminating flow, and an in-transit flow. We run simulation experiments in which the cell arrival process at the source of each flow is Bernoulli, and the arrival rate for each flow is 0.33 times the link data rate; hence, the traffic is admissible. In-transit and terminating flows are given weight 10 times larger than locally originating flows. Figure 2.4 shows that queue sizes exhibit a divergent

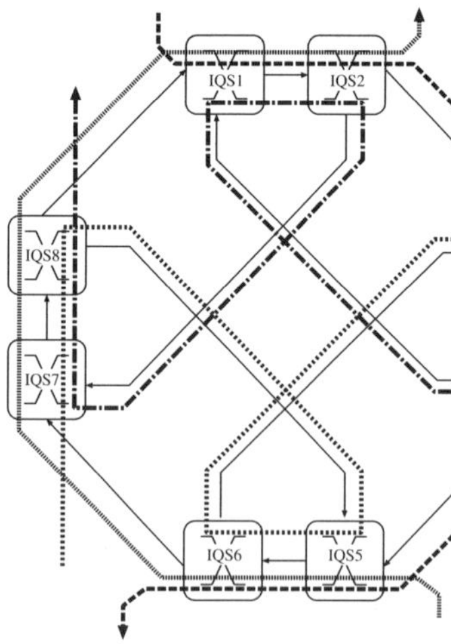

Figure 2.3. The network of IQ switches considered in our simulation

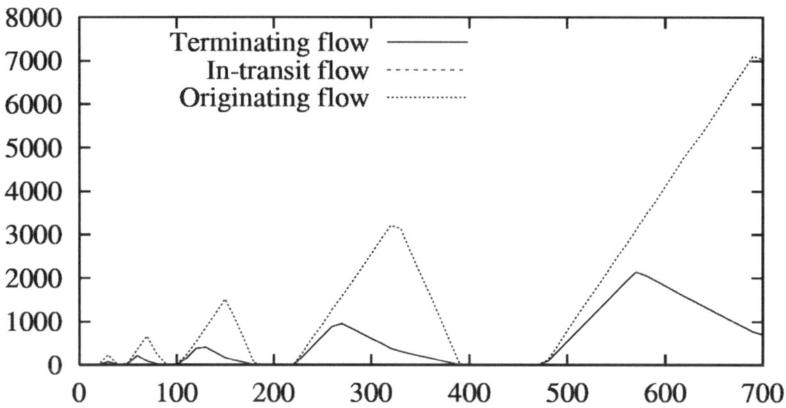

Figure 2.4. Queue sizes versus time in slots for the three flows at IQS1, when a local MWM is computed at each switch

oscillating behavior when a local MWM scheduling is adopted.

We show in the sequel how to modify the MWM to guarantee the maximum throughput in a network.

2.4.1 Theoretical Performance

We consider a network of K IQ switches. Switch k, $0 \leq k < K$, has P_k input ports and P_k output ports, all at the same cell rate. Each switch adopts VOQ at inputs. Thus there are P_k^2 different VOQs at switch k.

Thus, the network of switches can be modeled as a system Q containing $N = \sum_k P_k^2$ virtual queues. We restrict our study to the case $P_k = P \; \forall k$, so that $N = P^2 K$. Let $S(q)$ be the function that returns the switch on which VOQ q is located; let $I(q)$ be the function that returns the index of the input card at switch $S(q)$ on which the VOQ is located; let $O(q)$ be the function that returns the index of the output card at switch $S(q)$ to which VOQ cells are directed.

We adapt as follows the concept of $||Z||_{\mathrm{maxL}}$ to the case of a network of switches.

Definition 10 *Given a vector $Z \in \mathbb{R}^N$, $Z = \{z^{(n)}, n = P^2k + Pi + j,\ 0 \leq k < K,\ i, j = 0, 1, \ldots, P-1\}$, the norm $||Z||_{IO}$ is defined as*

$$||Z||_{IO} = \max_{\substack{k = 0, \ldots, K-1 \\ i = 0, \ldots, P-1}} \left\{ \sum_{n \in S^{-1}(k) \cap O^{-1}(i)} |z^{(n)}|, \sum_{n \in S^{-1}(k) \cap I^{-1}(i)} |z^{(n)}| \right\} \tag{2.33}$$

At each time slot, a set of non-contending cells departs from the VOQs of each switch. More formally, we say that, at each time slot, the departure vector $D \in \{0, 1\}^N$ must satisfy the condition:

$$||D||_{IO} \leq 1$$

The $N \times N$ matrix $R = [r^{(k,l)}]$ is the routing matrix: element $r^{(k,l)} = 1$ for cells departing from VOQ k and entering VOQ l.

Definition 11 *The traffic pattern loading a network of IQ switches is admissible if and only if*

$$||E||_{IO} = ||\Lambda(I - R)^{-1}||_{IO} \leq 1$$

Note that an admissible traffic pattern can be transferred without losses in a network of OQ switches.

The following results are taken from [3].

Theorem 7. *An open network of IQ switches implementing the $F(X)$-max-scalar policy is rate stable under each admissible traffic pattern such that arrival sequences at VOQs satisfy the strong law of large numbers, if:*

- $G(X) = F(X)[(I - R)^{-1}]^T$ *defines a conservative field;*
- $F(X)$ *satisfies conditions (2.21) and (2.22);*
- $F(\alpha X) = \alpha F(X)$ *for all scalars* α.

Similarly to the case of a single switch in isolation, for a network of switches it is possible to extend the result to more general functions $F(X)$ under any admissible i.i.d traffic pattern (*i.e.* under a smaller class of traffic patterns with respect to the assumptions of Theorem 7), by directly applying the Lyapunov function methodology to equations describing the stochastic evolution of the system.

Theorem 8. *An open network of IQ switches implementing the $F(X)$-max-scalar policy is strongly stable under each i.i.d. admissible traffic pattern, if:*

- $G(X) = F(X)[(I - R)^{-1}]^T$ *defines a conservative field;*
- $F(X)$ *satisfies conditions (2.21) and (2.22).*

As a consequence, the following result can be shown, corresponding to the policy proposed in [33].

Theorem 9. *An open network of IQ switches implementing the $F(X)$-max-scalar policy, with $F(X) = X(I-R)^{-1}$ is rate stable under each admissible traffic pattern such that the sequences of arrivals at VOQs satisfy the strong law of large numbers.*

The previous stable policy corresponds to a local MWM matching where the weight $w(q)$ associated to queue q of size $x(q)$ is given by

$$w(q) = \max\{0, x(q) - x(d[q])\}$$

when $d[q]$ is the downstream queue where cells from queue q are routed after being served[7].

The implementation of the $F(X)$-max-scalar policy can be performed in a distributed fashion; however, in this case, the implementation requires an exchange of information among neighbor switches. Some other possible solutions have been studied in [2].

2.5 Conclusions

In this chapter we have shown how the stochastic Lyapunov function methodology was employed during recent years as a powerful and versatile analytical tool to assess the performance of input queued switches. Thanks to this approach, researchers have been able to study throughput and/or delay performance not only for isolated switches, but also for networks of switches.

The brief discussion of the results reported in this chapter hides the main difficulty in employing this methodology; indeed, the design of a Lyapunov function, specific to the system under study and suited to prove its stability properties, requires creative effort, which is the most difficult step in the theoretical investigation. We strongly recommend the interested readers to refer to all the original papers and proofs in order to gain insight into the identification of the most suitable Lyapunov function for the system under study.

References

1. Ajmone Marsan M, Bianco A, Giaccone P, Leonardi E, Neri F (2002) Packet-mode scheduling in input-queued cell-based switches. *IEEE/ACM Transactions on Networking* 10:666–678
2. Ajmone Marsan M, Giaccone P, Leonardi E, Neri F (2003) On the stability of local scheduling policies in networks of packet switches with input queues. *IEEE JSAC* 21:642-655
3. Ajmone Marsan M, Leonardi E, Mellia M, Neri F (2005) On the Stability of Isolated and Interconnected Input-Queueing Switches Under Multiclass Traffic. *IEEE Transactions on Information Theory* 45:1167-1174
4. Anderson T, Owicki S, Saxe J, Thacker C (1993) High Speed Switch scheduling for local area networks. *ACM Transactions on Computer Systems* 319-352
5. Andrews M, Vojnovic M (2003) Scheduling reserved traffic in input-queued switches: new delay bounds via probabilistic techniques. *IEEE JSAC* 21:595-605
6. Andrews M, Zhang L (2001) Achieving stability in networks of input-queued switches. *Proc. of IEEE INFOCOM 2001* 1673-1679
7. Baskett F, Chandy KM, Muntz RR, Palacios F (1975) Open, closed and mixed networks with different classes of customers. *Journal of the ACM* 22:248-260

[7] For a proper definition of $w(q)$, it is defined $x(d[q]) = 0$ when q is an egress queue of the network.

8. Cheng-Shang C, Wen-Jyh C, Hsiang-Yi H (2000) Birkhoff-von Neumann input buffered crossbar switches. *Proc. of IEEE INFOCOM 2000*
9. Chuang ST, Goel A, McKeown N, Prabhakar B (1999) Matching output queuing with combined input and output queuing. *IEEE JSAC* 17:1030-1039
10. Dai JG (1999) Stability of Fluid and Stochastic Processing Networks. In: Miscellanea Publication n.9, Center for Mathematical Physics and Stochastic, Denmark (`http://www.maphysto.dk`)
11. Dai JG (1995) On Positive Harris Recurrence of Multiclass Queueing Networks: a Unified Approach Via Fluid Limit Models. *Annals of Applied Probability* 5:49-77
12. Dai JG, Prabhakar B (2000) The throughput of data switches with and without speedup. *Proc. of IEEE INFOCOM 2000* 556-564
13. Fayolle G (1989) On random walks arising in queueing systems: ergodicity and transience via quadratic forms as Lyapunov functions – Part I. *Queueing Systems* 5:167-184
14. Y.Ganjali Y, A.Keshavarzian A, D.Shah D (2005) Cell Switching Versus Packet Switching in Input-Queued Switches. *IEEE/ACM Transactions on Networking* 13:782-789
15. Giaccone P, Prabhakar B, Shah D (2003) Randomized scheduling algorithms for high-aggregate bandwidth switches. *IEEE JSAC* 21:546-559
16. Karol M, Hluchyj M, Morgan S (1987) Input versus output queuing on a space division switch. *IEEE Transactions on Communications* 35:1347-1356
17. Kleinrock L (1975) *Queuing Systems* – Vol. I: Theory, J. Wiley
18. Kumar PR, Meyn SP (1995) Stability of queueing networks and scheduling policies. *IEEE Transactions on Automatic Control* 40:251-260
19. Kushner HJ (1967) *Stochastic Stability and Control*, Academic Press
20. LaMaire RO, Serpanos DN (1994) Two dimensional round-robin schedulers for packet switches with multiple input queues. *IEEE/ACM Transactions on Networking* 2:471-482
21. Leonardi E, Mellia M, Neri F, Ajmone Marsan M (2001) On the stability of Input-Queued Switches with Speed-up. *IEEE/ACM Transactions on Networking* 9:104-118
22. Leonardi E, Mellia M, Neri F, Ajmone Marsan M (2003) Bounds on delays and queue lengths in input-queued cell switches. *Journal of the ACM* 50:520-550
23. McKeown N (1999) The iSLIP scheduling algorithm for input-queued switches. *IEEE Transactions on Networking* 7:188-201
24. McKeown N, Mekkittikul A (1998) A practical scheduling algorithm to achieve 100% throughput in input-queued switches. *Proc. of IEEE INFOCOM'98* 792-799
25. McKeown N, Anantharam V, Walrand J (1996) Achieving 100% Throughput in an Input-Queued Switch. *Proc. of IEEE INFOCOM'96* 296-302
26. McKeown N, Mekkittikul A, Anantharam V, Walrand J (1999) Achieving 100% throughput in an input-queued switch. *IEEE Transactions on Communications* 47:1260-1267
27. Neely NJ, Modiano E (2004) Logarithmic delay for NxN packet switches. *Proc. of IEEE HPSR 2004* 3-9
28. Ross K, Bambos N (2004) Local search scheduling algorithms for maximal throughput in packet switches. *Proc. of IEEE INFOCOM 2004* 2:1158-1169
29. Shah D, Kopikare M (2002) Delay Bounds for Approximate Maximum Weight Matching Algorithms for Input Queued Switches. *Proc. of IEEE INFOCOM 2002*
30. Tamir Y, Frazier G (1988) High performance multi-queue buffers for VLSI communication switches. *Proc. of 15th Annual Symposium on Computer Architecture* 343-354
31. Tamir Y, Chi HC, Symmetric crossbar arbiters for VLSI communication switches (1993). *IEEE Transactions on Parallel and Distributed Systems* 4:13-27
32. Tarjan RE (1983) Data structures and network algorithms, Society for Industrial and Applied Mathematics, Pennsylvania

33. Tassiulas L, Ephremides A (1992) Stability properties of constrained queueing systems and scheduling policies for maximum throughput in multihop radio networks. *IEEE Transactions on Automatic Control* 37:1936-1948
34. Tassiulas L (1998) Linear complexity algorithms for maximum throughput in radio networks and input queued switches. *Proc. of IEEE INFOCOM'98* 553-559

3

Adaptive Batch Scheduling for Packet Switching with Delays

Kevin Ross[1] and Nicholas Bambos[2]

[1] University of California Santa Cruz, kross@soe.usc.edu
[2] Stanford University, bambos@stanford.edu

We discuss the control of a packet switch where delays due to mode switching become important. Whereas most packet switch scheduling analysis assumes that switches can operate with negligible delays, we consider what to do when this does not hold.

Several practical situations can indeed lead to a time lag in switches, where changing physical connections in the switch can take a significant time relative to the high data rates processed in the fabric. If switches are forced to change modes too frequently, this leads to a loss of throughput.

This chapter provides a synthesis of recent developments in packet switch scheduling with delays. We present an overview of how to manage a switch when there are forced delays at the switch; that is, when the required time to change from one switching mode to another is nontrivial. If traditional policies such as weighting-based or online fixed batch sizes are utilized, there can be a loss of throughput. We will discuss the result that *adaptive batch scheduling algorithms* maintain maximal throughput in a switch *even in the presence of nontrivial delays*.

3.1 Introduction

Switches provide the connections between network links via various architectures, enabling packets of data to provide communication over the internet. A packet arriving at the switch is stored in a buffer until connections in the switch allow it to traverse toward its chosen destination.

Particularly in modern switches, *several* incoming lines are routed to *several* outgoing lines in the same switch. For example, the crossbar design of many switches allows multiple simultaneous connections. Each input port can be connected to exactly one output port, and each output port accepts connections from exactly one input. Thus in a two-by-two switch, there are two modes available, connecting input–output pairs $(1,1)$ and $(2,2)$ or $(1,2)$ and $(2,1)$. This is illustrated for the case of two incoming and outgoing lines in Figure 3.1, where the switch is set to one of its two available modes.

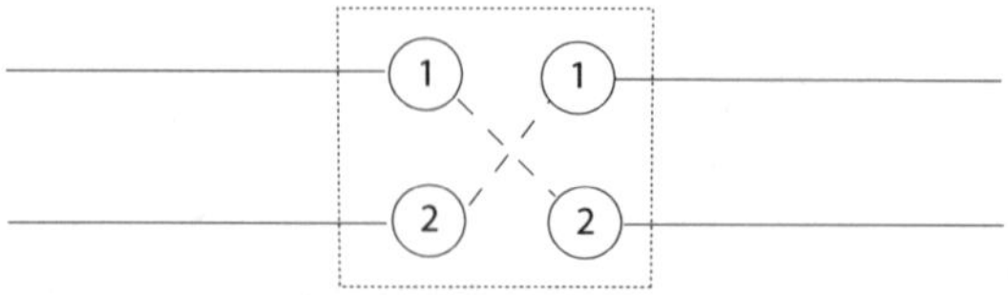

Figure 3.1. A 2-by-2 crossbar switch. A crossbar switch connects multiple input to output pairs, and can connect more than one pair at a time. This figure shows a 2-by-2 switch with two input ports and two output ports. At any time, the crossbar architecture in the switch allows each port to be connected to exactly one other port. There are two modes available, connecting the pairs (1,1) and (2,2) or the pairs (1,2) and (2,1), as depicted above.

Modern switches also include features such as speedup, multicast capability and class differentiation. All of these can be described via the model in this chapter, but we focus this study on the particular issue of switch delays. The type of switch defines the options available for processing packets.

Several recent studies have proposed algorithms for efficient packet scheduling in high-performance networks (see, for example [1–4]). An important feature absent from many of these analytical studies is the problem of switch delay. When a switch transfers packets via one particular operational mode, it may not be possible to instantaneously shift to a new mode. The ability to do this depends upon the *switch fabric* and the *scale* of the switch.

Typically, if the delay between modes is ignored, switching policies tend to operate with *frequent* mode changes. Algorithms such as *matching*-based policies draw the switch toward a balanced load where many modes are equally desirable. While this has the advantage of load balancing the switch while maintaining stability, the frequent mode changes can be highly inefficient. For example, if a switch spends half of its time in each of two modes m_1 and m_2, the sequence $\{m_1, m_2, m_1, m_2, m_1, m_2, m_1, m_2, ...\}$ causes twice as much aggregate delay as the sequence $\{m_1, m_1, m_2, m_2, m_1, m_1, m_2, m_2, ...\}$. In a fully utilized switch, this actually **reduces switch throughput**, as will be explained further in this chapter.

3.2 Switching Modes with Delays: A General Model

Our model describing a general switch is illustrated in Figure 3.2, with Q queues and M service modes to select from. The queues are labeled $q \in \mathcal{Q} = \{1, 2, ..., Q\}$, with each queue storing a unique type of packet.

We utilize *virtual output queues*, illustrated for a 2-by-2 switch in Figure 3.3. Instead of having one buffer for each input port, the input port buffer is separated into parallel queues, one for each output port. Thus there is a queue for each input–output pair, avoiding potential problems such as head of line blocking [5] and keeping a clear management of queue levels.

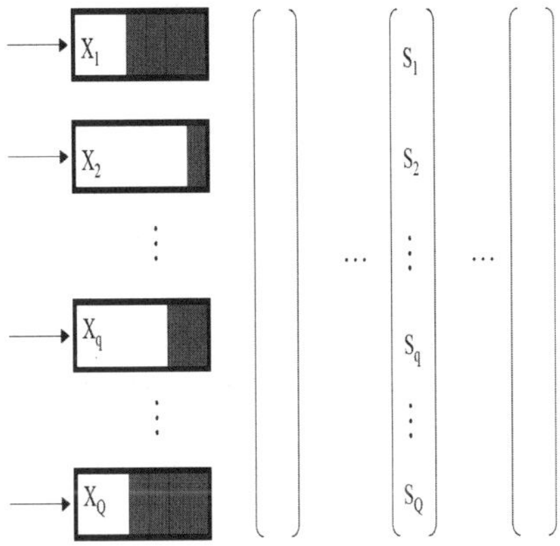

Figure 3.2. The core model is illustrated above. Any switch can be described as cells waiting in a set of parallel queues, and a finite set of available service modes or switch modes to choose from. The selected mode describes the set of rates at which all of the queues are served under that mode. In most switches (without speedup), this is a set of zero and one entries.

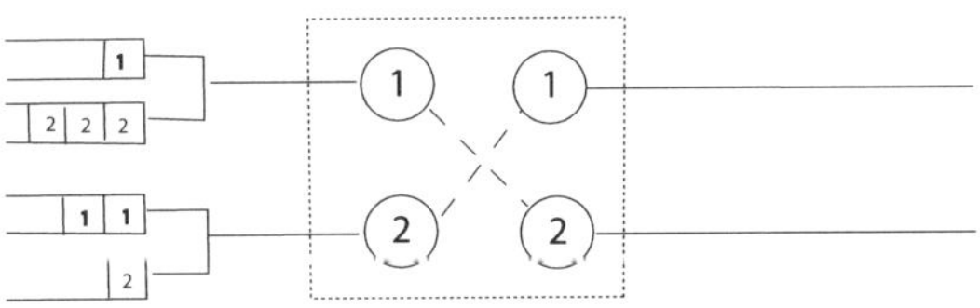

Figure 3.3. Virtual output queues. To avoid the potential effect of head-of-line blocking, each input–output pair has its own queue. Packets within that queue can be served first-in-first-out (FIFO) or according to some other discipline, without causing any unnecessary starvation effects.

Time is divided into equal length time slots, labeled $t \in \{0, 1, 2, ...\}$. We will consider the dynamics of the switch in terms of these discrete units of time, but note they can be arbitrarily small.

For the sake of simplicity, we will assume that a data packet can be divided into units of workload called cells. $A_q(t)$ is the number of cells arriving at queue q in time slot t.

For this model, we describe everything in terms of Q-length vectors. The individual elements of the vector correspond to each queue. For example, the arrival vector in time slot t is $A(t) = (A_1(t), A_2(t), ..., A_q(t), ..., A_Q(t))$.

Throughout this chapter we will refer to the vectors, mentioning individual subscript q only when necessary.

At any time slot, the switch may be in one service mode m, chosen from a *finite* set $\mathcal{M}$={1,2,...,M}. Each mode m has a corresponding Q-dimensional vector $S^m = (S_1^m, S_2^m, ..., S_q^m, ..., S_Q^m)$, where $S_q^m \geq 0$ is the number of cells removed from queue q in a single time slot, when the switch operates under mode m. Arriving cells are buffered immediately in their respective queues, and served in a first-in-first-out (FIFO) manner within each queue.

Separating this analysis from most previous work, the mode selection is *sequence dependent*. Two different modes cannot necessarily be used in consecutive time slots. This takes account of many complexities in switches, including the case where the time slots are small enough that switches cannot physically reconfigure as fast as the packets can be transferred.

For any pair of distinct service modes, m_1 and m_2, δ is the required down time from the activation of the first mode until the second can be activated. If mode m_1 is used at time t then m_2 may not be used in the interval $(t, t+\delta)$. In particular, if m_1 and m_2 are to be used consecutively then there can be no active service mode in time slots $(t+1, t+2, ..., t+\delta)$.

Let $m = 0$ denote the mode where no service is applied to any queue, corresponding to $S^0 = 0$, the zero vector. The scheduling problem is to select the service mode $m(t)$ for each time slot with the added constraint:

$$m(t+r) \in \{m(t), 0\} \text{ for } r \in \{1, 2, ..., \delta\} \tag{3.1}$$

In a related problem, one could consider nonuniform delays δ_{m_i,m_j} between modes. The analysis of throughput in this chapter still holds by recognizing that the longest required delay $\delta_{\max} = \max_{m_i,m_j \in \mathcal{M}} \delta_{m_i,m_j}$ must be bounded, but for simplicity in the exposition we will discuss the case of uniform delays where $\delta_{m_i,m_j} = \delta$.

The backlog state $X(t)$ of the queueing switch at any time t is the Q-dimensional vector of cell populations in the individual queues, that is,

$$X(t) = (X_1(t), X_2(t), ..., X_q(t), ..., X_Q(t)) \tag{3.2}$$

where $X_q(t)$ is total workload (number of waiting cells) existing in queue $q \in \mathcal{Q}$ at time $t \in \{0, 1, 2, ...\}$. The **service state** $S(t)$ of the switch at time t is the service vector

$$S(t) = S^{m(t)} = (S_1(t), S_2(t), ..., S_q(t), ..., S_Q(t)) \tag{3.3}$$

chosen by the scheduling algorithm that the switch operates under via the selection of mode $m(t)$ at time t. Clearly, $S(t) = 0$, the zero vector when mode $m(t) = 0$ is selected.

The cell backlog in the switch at time t can be expressed as

$$\begin{aligned} X(t) &= [X(t-1) - S(t-1)]^+ + A(t-1) \\ &= X(0) + \textstyle\sum_{s=0}^{t-1} A(s) - \sum_{s=0}^{t-1} \hat{S}(s) \end{aligned} \tag{3.4}$$

for $t \in \{1, 2, ...\}$ with $X(0)$ the starting backlog, and $\hat{S}_q(t) = \min\{X_q(t), S_q(t)\}$ ensures that no queue has negative workload in any time slot. This assumes a store-and-forward switch, where packets cannot arrive and be processed in the same slot.

The alternative, a cut-through switch, could also be described by adjusting the evolution equations appropriately.

For each $q \in \mathcal{Q}$ we assume that

$$\lim_{t\to\infty} \frac{\sum_{s=0}^{t-1} A_q(s)}{t} = \rho_q \in (0, \infty) \tag{3.5}$$

that is, the long-term arriving traffic load for each queue q is well defined, positive and finite. We make no further restriction on the traffic traces. We do not assume any particular statistics that may generate the traffic traces. This means that we allow arrivals to different queues to be correlated and even directly dependent.

For a switch to be *rate stable*, we require

$$\lim_{t\to\infty} \frac{\sum_{s=0}^{t-1} S_q(s)}{t} = \lim_{t\to\infty} \frac{\sum_{s=0}^{t-1} A_q(s)}{t} = \rho_q \tag{3.6}$$

for each $q \in \mathcal{Q}$. One can see from (3.4) and (3.6) that for such rate stability it suffices to show

$$\lim_{t\to\infty} \frac{X(t)}{t} = 0 \tag{3.7}$$

that is, the workload in each queue is not growing linearly over time. Recall from (3.1) that some of the time slots must have $S(t) = 0$, so the ability to guarantee (3.7) is not immediately clear.

The set of Q-vectors ρ for which there exists a sequence of $S(t)$ vectors which would give rate stability and hence satisfy (3.6) is known as the *stability region* $\mathcal{R}$. A rate vector is called allowable if it is within the stability region. The stability region $\mathcal{R}$ is the set of allowable rate vectors ρ satisfying

$$\mathcal{R} = \left\{ \rho : \ \rho \le \sum_{m=1}^{M} \phi_m S^m \ \text{ for some } \phi_m \ge 0, \ \sum_{m=1}^{M} \phi_m = 1 \right\} \tag{3.8}$$

In simple terms, a workload arrival rate is allowable, or *in the stability region* if there exists a convex combination of available service modes which would serve at least the arrival rate for every queue. If one knew ρ in advance, an algorithm which used each S^m for the fraction ϕ_m of time and somehow minimized the time spent in transition mode $S^0 = 0$ would ensure that (3.6) was satisfied. Here, we are concerned with the case where ρ is not known, and a scheduling algorithm allocates the switch mode according to observed backlog.

3.3 Batch Scheduling Algorithms

Whereas most work on maximal throughput algorithms for packet switches considers the online scheduling of each mode on a per time slot basis [1–4,6,7], here we present **batch** policies, where several consecutive time slots are scheduled in a group.

Batch policies are a natural approach in the presence of delays, since they take a longer term perspective on scheduling. A sequence of packets are allowed to arrive and form an aggregate group, or batch. This batch of packets are scheduled together to be transferred through the switch in subsequent time slots. Since the batch are scheduled together, the effect delays between mode changes can be reduced by scheduling consecutive slots with the same mode instead of alternating between modes too frequently. We present two families of batch policies, with fixed or adaptive batch lengths, and argue that the fluctuations in packet switch load make it often preferable to utilize the adaptive policy.

3.3.1 Fixed Batch Policies

In most approaches that consider batches, such as round-robin based schemes, the size of each batch is *fixed*. That is, the scheduling is done every say T time slots, for some appropriate batch size T. In the extreme case, and indeed for many algorithms, $T = 1$ is assumed, scheduling just one time slot at a time. More general algorithms schedule *batches* of modes, as seen in the examples here. We call these policies/algorithms *fixed batch schedules* (FBS).

The utilization of a batch-based policy mitigates *some* of the effect of delays between modes. For example, a batch might instruct the switch to use mode m_1 for five time slots, and then mode m_2 for three time slots, with only one transition required from mode m_1 to mode m_2. A per time slot algorithm might use the same combination of modes but in a different order, leading to more mode transitions and therefore more down time.

Let us consider how the workload evolution would happen under an FBS policy. We assume the existence of a static scheduling policy Σ, that schedules a sequence of modes and their associated delays over T time slots based on the waiting workload X at the start of the batch. If it is possible to schedule all waiting packets at T, Σ will do so, otherwise it will schedule a sequence of modes to process a subset of the waiting packets.

Let $T_0 = 0$, and inductively define for $N = 0, 1, 2, 3, ...$ the formation of batches in FBS and their processing intervals, as follows.

1. At time $T_N = N.T$, schedule the batch demand $X(T_N)$ of packets waiting in the switch at time T_N, according to the static schedule Σ. The workload waiting in $X(T_N)$ is processed in the time interval $[T_N + 1, T_N + T]$ according to Σ.
2. New packets arriving throughout the time interval $[T_N + 1, T_N + T]$ are not processed, but stored until time $T_{N+1} = T_N + T$. Thus, at time T_{N+1} a new batch $X(T_{N+1})$ has been formed by accumulating the packets that arrived after T_N and up to time T_{N+1}, plus any packets that were not completely served by Σ in the previous batch. The batch $X(T_{N+1})$ is then scheduled at T_{N+1} according to the static schedule Σ and the process repeats itself.

The fixed batch schedules are simple to implement, especially if Σ is easy to calculate. There are two notable disadvantages, however, of the fixed batch policies.

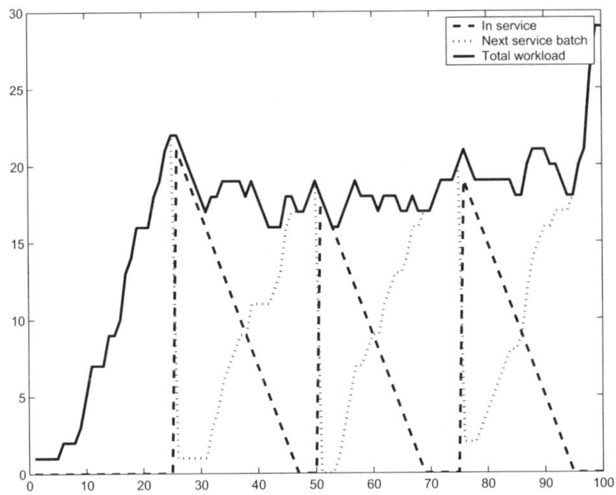

Figure 3.4. Fixed batch scheduling. The sample trace of a fixed batch scheduling algorithm is shown. The workload in the vertical axis is recorded from three perspectives: (i) what the server sees, (ii) what the server cannot see, and (iii) the total workload waiting.

Namely (i) the batches are not responsive to bursts in the number and size of packets. This means that if there is a relatively small backlog at the start of a batch, it will still take T time slots to process before the new arriving packets can be served, and (ii) fixed batches cannot guarantee maximal throughput for fixed T.

Consider briefly why fixed batches cannot guarantee maximal throughput. In any given batch, there is at least some down time $\delta > 0$, which varies based on the schedule created by Σ. This follows since each batch is independent of the previous batch, so there must be some delay between consecutive batches (as well as any delays within the schedule itself). Thus the switch is utilized for a fraction of at most $\frac{T-\delta}{T}$ of the batch, and the throughput capacity is reduced by that fraction. In general the reduction will be much more than that, since multiple switching times are usually required in a given batch.

One could make the argument that for any $\rho \in \mathcal{R}$, there exists a large enough batch size T_ρ to maintain rate stability. However, the implementation of such a batch policy would then require advance knowledge or ρ in order to fix the appropriate batch length. Further, such an argument exemplifies the other disadvantage above, namely *large batches can lead to large inefficiencies*.

The workload evolution for FBS is illustrated in Figure 3.4, recording the total workload as well as the workload in the current batch and the arriving workload for the next batch. It is particularly informative to note that during the first batch, when nothing is being processed, the workload grows very quickly.

3.3.2 Adaptive Batch Policies

We contrast the FBS algorithms with a class of adaptive batch scheduling (ABS) algorithms. In ABS, the batches are of different length, depending on the demand waiting at the switch when the batch begins.

In a similar manner to the fixed batch schedule, each batch calculates a *static* schedule based on the demand (backlog), but here the static batch is run to completion. During a static batch, new arrivals add workload to the switch. At the end of each batch a new schedule is set based on the workload that arrived during the previous batch. If there is a lot of workload waiting, the schedule will use more time than if a small workload is to be processed.

Suppose workload vector X is given and also some *static* feasible processing schedule Σ (possibly suboptimal), which processes the demand X in $\tau(X)$ time slots. Σ corresponds to a sequence of service modes $m(t)$, with appropriate down time between changes inserted to satisfy the condition in (3.1).

Individual static batch schedules are considered in more detail later, but first consider how Σ would be used to process dynamically arriving data packets. The schedule is referred to as a static schedule since it considers the fixed demand at the start of the batch, and activates a schedule for subsequent time slots. Just as in FBS, the batch schedule does not adjust with new arriving packets.

Starting at time $T_0 = 0$, inductively define for $N \in \{1, 2, 3, ...\}$ the formation of batches, as follows.

1. At time T_N schedule the batch demand $X(T_N) \neq 0$ of packets waiting in the queues at T_N, according to the static schedule Σ, which clears it at time $T_{N+1} = T_N + \tau(X(T_N))$. Packets in $X(T_N)$ are processed in the time interval $[T_N + 1, T_N + \tau(X(T_N))]$.
2. New packets arriving throughout the time interval $[T_N + 1, T_N + \tau(X(T_N))]$ are not processed, but stored until time $T_{N+1} = T_N + \tau(X(T_N))$. Thus, at time T_{N+1} the previous batch $X(T_N)$ has been completely cleared, and a new batch $X(T_{N+1})$ has been formed by accumulating the packets that arrived after T_N and up to time T_{N+1}. The batch $X(T_{N+1})$ is then scheduled at T_{N+1} according to the static schedule Σ and the process repeats itself.
3. If no new packet arrives while a batch is served, or $X(T_{N+1}) = 0$, simply increment by default the time register $T_{N+2} = T_{N+1} + 1$ by one time slot.

According to the above scheme, the evolution of the increasing unbounded time sequence $\{T_N, N = 1, 2, 3, ...\}$ is given by the following induction formula:

$$T_{N+1} = \begin{cases} T_N + \tau(X(T_N)), & \text{if } X(T_N) \neq 0 \\ T_N + 1, & \text{if } X(T_N) = 0 \end{cases} \tag{3.9}$$

Note that the Nth batch $X(T_N)$ is comprised of the workload that arrives throughout the interval $[T_{N-1}+1, T_N]$, and is processed throughout the interval $[T_N+1, T_{N+1}]$.

It is naturally possible to utilize excess capacity by processing new arrivals in a previous batch. For example, a batch may schedule the switch to process a single

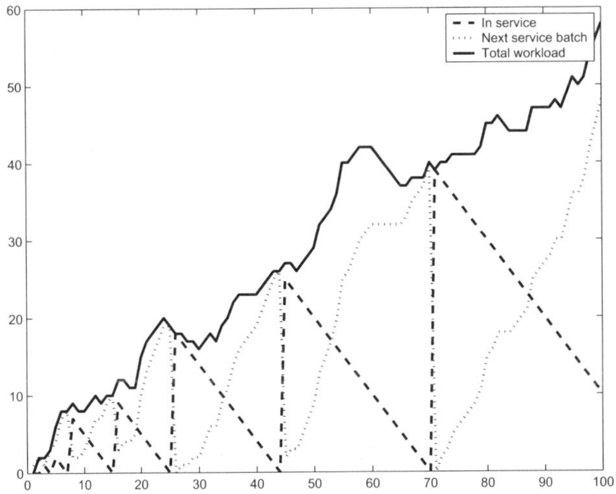

Figure 3.5. Adaptive Batch Scheduling. The sample trace of an adaptive batch scheduling algorithm. The workload in the vertical axis is recorded from three perspectives, (i) what the server sees, (ii) what the server cannot see and (iii) the total workload waiting. This differs from the fixed batch policy, where each batch is the same length.

queue when two could be served in parallel without additional time, if there were workload in the second queue. If packets arrive at the parallel queue before the batch begins processing the packets, they can be processed before the next batch is scheduled. We ignore these packets for the sake of analysis, except to note that this will only *improve* the performance of the switch.

The trace of workload under an ABS algorithm is illustrated in Figure 3.5. The batch lengths increase as the workload increases, and small batches are used when the workload is small. The figure contrasts ABS with the case of fixed batch lengths, which do not necessarily serve all the workload waiting for a single batch.

3.3.3 The Simple-batch Static Schedule

Here we focus on a static schedule known as simple-batch. The simple-batch policy is based on the solution to the following linear program:

$$\min \sum_{m=1}^{M} \alpha^m \text{ such that } \sum_{m=1}^{M} \alpha^m S^m \geq X, \alpha^m \geq 0 \tag{3.10}$$

A feasible solution to (3.10) is a sequence $\{\alpha^m\}$ where α^m represents the time to spend in mode S^m in order to clear the workload X. The optimal solution clears the demand in minimum time if there were no transition time restrictions from (3.1). The solution to the linear program in (3.10) has at most Q nonzero α^m values, corresponding to the set $\{m_1, m_2, ..., m_Q\}$. Since delays are uniform this corresponds to a total of $Q\delta$ delay within the batch. One could consider a more complicated problem, optimizing over delays as well to make a potentially more efficient schedule. Unfortunately, this would no longer be a strict *linear* program, since we would require binary variables to recognize which modes were included. Fortunately, however, (3.10) turns out to be sufficient for stability, as seen in Section 3.5.

The *simple batch* policy solves the linear program in (3.10), and puts the service vectors corresponding to nonzero α^m values in some order $\{m_1, m_2, ..., m_Q\}$, and inserts the necessary waiting time between modes. The modes are then scheduled as follows:

1. Set $k = 1$.
2. Use mode m_k for α^{m_k} time slots.
3. Wait δ time slots.
4. $k = k + 1$.
   If $k = Q$ then wait δ and stop, else go to step 2.

It is easily seen that the simple batch schedule meets the demand X and respects the necessary transition requirements, since it uses a solution to (3.10). This static schedule could be utilized in either FBS or ABS. Under ABS it would be run to completion, whereas under FBS the schedule would either be truncated at the fixed time T, or the schedule would need to apply an additional constraint, scaling down the total time used so that $\sum_m \alpha^m + Q\delta \leq T$.

The ordering of the service modes could be chosen according to a priority rule or optimized for performance. For example, service to a priority queue could be scheduled first within the sequence of modes selected by (3.10).

The simple-batch schedule has a total busy processing time of $\sum_m \alpha^{m_*}$ and maximum total transition time of $Q\delta$. Recall from (3.8) that by stability $\rho \leq \sum_{m=1}^M \sigma^m S^m$ for some $\{\sigma^m\}_{m=1}^M$ with $\sum_{m=1}^M \sigma^m = 1$. Given this, for any scalar value $\beta > 0$ and workload vector $\beta\rho$, a multiple of that scalar, it follows that $\tau(\beta\rho) \leq \beta + Q\delta$.

3.4 An Interesting Application: Optical Networks

With the general switching model described in this work, a surprising number of applications can be controlled using the switch scheduling algorithms. For example, networks of optical fibers over large geographical areas have been modeled as a generalized *switch* since they involve dynamic scheduling of cross-connects.

The application arises where packets are transmitted among a multiplicity of geographically separated source and destination nodes. Here the delay between modes is required due to discrepancies in the distance traversed by each data packet. It occurs in a recently developed wavelength-division multiplexed (WDM) transport network, called *Time-domain Wavelength Interleaved Network* (TWIN) [8–10]. In TWIN the queues correspond to pairs of nodes or cities, and the service configurations to the combination of connections which can be simultaneously established. Just as in a

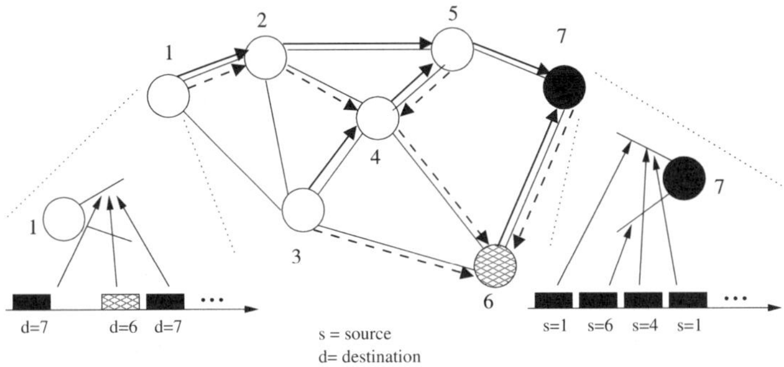

Figure 3.6. TWIN network. Logical multipoint-to-point trees (one shown for each of the two destinations), are overlayed on top of a physical topology. Node 1 interleaves its transmissions to nodes 6 and 7 by tuning its laser to λ_6 when it wants to transmit a packet to node 6 and to λ_7 when it wants to transmit a packet to node 7. Each internal node in the network simply performs self-routing of packets based on their colors (*i.e.* wavelengths). For example, node 2 will route an incoming packet of λ_6 to node 4 and an incoming packet of λ_7 to node 5.

crossbar switch, each port can connect to one other port, each city in an optical network can communicate with one other via the fiber-optic network.

TWIN defines an optical network with a simple core architecture consisting of wavelength-selective cross-connects capable of routing incoming data packets to the appropriate outgoing ports or fibers. The cross-connect configuration in the core is fixed among source and destination nodes. Fast tunable lasers at the source nodes send streams of data at frequencies corresponding to the destination nodes. In TWIN, a unique wavelength λ_j is assigned to each destination j, and sources that have data to transmit to a particular destination use the wavelength assigned to that destination.

Due to this fixed-wavelength policy, the wavelength connections from various sources to a particular destination j can be viewed as an optical multipoint-to-point connection of wavelength λ_j rooted at the destination j. To ensure that packets of the same wavelength do not arrive at the same destination simultaneously, sources must coordinate their transmissions over time. TWIN relies on scheduling to avoid such destination conflicts.

Figure 3.6 shows a simple TWIN architecture where multipoint-to-point optical connections rooted at two destinations (nodes 6 and 7) have been configured. Typically, reconfigurations occur at a relatively long time scale (*e.g.* hours, days or weeks), so the scheduling is performed on a fixed network connection configuration. The individual packet forwarding based on assignment of appropriate colors at the source is performed at a relatively small time scale (microseconds).

Time slots are such that each node can send or receive exactly one packet. At each time slot, the scheduler needs to assign the appropriate wavelength to each tunable laser. The scheduler must avoid collisions at destinations and make efficient use of the time slots.

Consider a network of C nodes (representing cities) and denote their set by $\mathcal{C} = \{1, 2, 3, ..., C\}$. For each communicating source–destination pair $(i, j) \in \mathcal{C} \times \mathcal{C}$ assign a queue, where the data packets originating in node i and destined to node j are queued up while waiting to be transmitted. The queues here are labeled by the source (i) and destination (j) nodes which they correspond to.

Assume tree-based routing, which means that internal conflicts in the network are avoided by scheduling only at the source and destination nodes. For each source–destination pair (i, j) assume that there is a fixed, known propagation delay d_{ij} for sending a data packet from node i to j. Hence, a packet sent at time t from i to j will arrive at its destination at time $t + d_{ij}$. Assume for simplicity in the exposition that the propagation delay is an integer multiple of time slots.

Because of the different propagation delay values d_{ij} and the single laser and receiver at each node, the service configurations must satisfy the interleaving property

$$\sum_j S_{ij}(t) \leq 1, \ \forall i, \ \ \sum_i S_{ij}(t - d_{ij}) \leq 1, \ \forall j \tag{3.11}$$

for all times slots t.

Putting this in terms of the down time requirement of this chapter, one way to satisfy the propagation delays is to ensure

$$\delta_{m_1, m_2} = \max\{(d_{ij} - d_{i'j}) : S_{ij}^{m_1} = 1, S_{i'j}^{m_2} = 1 \text{ for some } i, i'\} \tag{3.12}$$

and use $\delta = \max_{m_1, m_2 \in \mathcal{M}} \{\delta_{m1, m_2}\}$ as the common delay.

It turns out to be sufficient to consider modes without conflicts at the sender or receiver, that is modes m with service vector S^m satisfying $\sum_i S_{ij}^m \leq 1, \sum_j S_{ij}^m \leq 1$, a subset of those satisfying (3.11).

The simple-batch policy in these networks turns out to be easily found using the Birkhoff–von Neumann decomposition [11], and the minimum schedule time is seen to be the maximum sending or receiving node demand. That is,

$$\tau^* = \max \left\{ \max_j \sum_i X_{ij}, \max_i \sum_j X_{ij} \right\} + Q\delta \tag{3.13}$$

This example of scheduling for TWIN is interesting since it utilizes the idea of a network being a *meta-switch*. Simpler examples of delays in switches also occur in large-scale modern routers, where physical connections need to be reconfigured under high data rates.

3.5 Throughput Maximization via Adaptive Batch Schedules

Here we outline the proof of throughput maximization for adaptive batch schedules; we both (i) identify conditions on which a static batch scheme leads to rate stability and (ii) show that the simple-batch policy satisfies those conditions. A full proof of this result appears in [12], and we sketch the main points here.

Recall from (3.6) that it is sufficient for rate stability to show $\lim_{t\to\infty}\frac{X(t)}{t}=0$, so this is the object of the stability proof. First we identify the conditions for a service policy $\{S(t)\}_{t=1}^{\infty}$ operating under ABS to induce rate stability. The properties are conditions on the static batch schedule Σ that is used to schedule packets for transmission within each individual batch. We then show that the simple-batch policy satisfies those conditions.

Since the demand $X(T_N)$ of the Nth batch is comprised of the packets that arrive in the time interval $[T_{N-1}+1, T_N]$, it follows that $X(T_N)=\sum_{t=T_{N-1}+1}^{T_N}A(t)$. This demand will be cleared by Σ at time $T_{N+1}=T_N+\tau(X(T_N))$, thus the choice of the static schedule Σ is the key to guaranteeing stability.

Here we outline the key steps of the proof the ABS leads to maximal throughput. A rigorous proof requires the consideration of cases where the limits are not well defined. In that case, one must consider subsequences of batches on which extreme limits are found. For more details, see [9, 12].

It turns out that the following condition on batch lengths is sufficient for rate stability:

$$\lim_{N\to\infty}\frac{T_N-T_{N-1}}{T_N}=0 \tag{3.14}$$

That is, (3.14) implies (3.6), so implies rate stability.

Outline of justification for (3.14)

Note that (3.14) implies $\lim_{N\to\infty}\frac{T_{N-1}}{T_N}=1$. Further, due to the structure of each batch, $X(T_N)=\sum_{t=T_{N-1}+1}^{T_N}A(t)$. Therefore, since $\{T_N\}_{N=1}^{\infty}$ is an increasing and unbounded sequence,

$$\begin{aligned}\lim_{N\to\infty}\frac{X(T_N)}{T_N} &= \lim_{N\to\infty}\frac{\sum_{t=T_{N-1}+1}^{T_N}A(t)}{T_N}\\ &= \lim_{N\to\infty}\frac{\sum_{t=1}^{T_N}A(t)}{T_N}-\lim_{N\to\infty}\frac{\sum_{t=1}^{T_{N-1}}A(t)}{T_{N-1}}\cdot\frac{T_{N-1}}{T_N}\\ &= \rho-\rho.1\\ &= 0\end{aligned} \tag{3.15}$$

where the above limit calculations hold for each individual component of the vectors. This implies [3] that $\lim_{t\to\infty}\frac{X(t)}{t}=0$, which is sufficient for rate stability by (3.6).

[3] Given any time slot $t\in\{1,2,3,...\}$, let t fall in the $N(t)$th batch interval, that is, $T_{N(t)}<t\le T_{N(t)+1}$. Consider the packets backlogged in the switch at time slot t. Each packet is included in one either $X(T_{N(t)})$ or $X(T_{N(t)+1})$. Therefore,

$$\begin{aligned}\lim_{t\to\infty}\frac{X(t)}{t} &\le \lim_{t\to\infty}\frac{X(T_{N(t)})+X(T_{N(t)+1})}{t}\\ &\le \lim_{t\to\infty}\frac{X(T_{N(t)})}{T_{N(t)}}+\lim_{t\to\infty}\frac{X(T_{N(t)+1})}{T_{N(t)+1}}\cdot\frac{T_{N(t)+1}}{T_{N(t)}}\\ &\le 0+0.1\\ &= 0\end{aligned} \tag{3.16}$$

A property of Σ, which is equivalent to (3.14) is

$$\lim_{N\to\infty} \frac{T_{N+1} - T_N}{T_N - T_{N-1}} \leq 1 \tag{3.17}$$

as long as each batch includes at most a bounded down time of Δ.

In other words, the length of a batch cannot grow linearly in time. Growing batches would indicate that the backlog in the queues was growing, since there is a direct relationship between batch length and initial workload.

To see this, we argue by contradiction[4], supposing that (3.17) holds and $\lim_{N\to\infty} \frac{T_N - T_{N-1}}{T_N} = \epsilon > 0$.

Note that $\frac{T_{N-1}}{T_N} \to 1 - \epsilon$.

$$\begin{aligned}
\epsilon &= \lim_{N\to\infty} \frac{T_N - T_{N-1}}{T_N} \\
&= \lim_{N\to\infty} \frac{T_N - T_{N-1}}{T_{N-1} - T_{N-2}} \frac{T_{N-1} - T_{N-2}}{T_{N-1}} \frac{T_{N-1}}{T_N} \\
&\leq 1.\epsilon.(1-\epsilon) \\
&< \epsilon
\end{aligned} \tag{3.18}$$

This is a contradiction, and completes the argument. Hence (3.17) establishes conditions on the static batch which lead to maximum throughput, and it remains to construct the static bach policies Σ which ensure that (3.17) is satisfied.

This can easily be shown for the simple-batch policy we have introduced. Noticing that $\lim_{N\to\infty} \frac{\sum_{t=T_{N-1}+1}^{T_N} A(t)}{T_N - T_{N-1}} = \rho$, we have

$$\begin{aligned}
\lim_{N\to\infty} \frac{T_{N+1} - T_N}{T_N - T_{N-1}} &= \lim_{N\to\infty} \frac{\tau(X(T_N))}{T_N - T_{N-1}} \\
&= \lim_{N\to\infty} \frac{\tau(\sum_{t=T_{N-1}+1}^{T_N} A(t))}{T_N - T_{N-1}} \\
&= \lim_{N\to\infty} \frac{\tau((T_N - T_{N-1})\rho)}{T_N - T_{N-1}} \\
&\leq \lim_{N\to\infty} \frac{T_N - T_{N-1} + Q\delta}{T_N - T_{N-1}} \\
&= 1
\end{aligned} \tag{3.19}$$

This implies (3.17) and hence rate stability is established.

3.6 Summary

We have outlined scheduling policies to guarantee maximal throughput in packet switching networks where delays are significant. The increasing data speeds in networks make it unrealistic to utilize scheduling algorithms that allocate switch modes and reconfigure physical devices at the same rate that data is arriving at each of the switch ports, especially when we consider applications such as the meta-switch network model.

[4] Again, we assume that all limits are well-defined. If this is not the case, one can take lim sup subsequences, as seen in [9, 12]

Adaptive batch scheduling algorithms mitigate the natural disadvantages of fixed batch algorithms, particularly their responsiveness to workload fluctuations and the ability to guarantee maximal throughput without prior knowledge of the arrival dynamics. ABS require the solving of a simple linear program at the beginning of each batch.

References

1. N. McKeown, A. Mekkittikul, V. Anantharam, and J. Walrand. Achieving 100% throughput in an input-queued switch. *IEEE Transactions on Communications*, 47(8):1260–1267, 1999.
2. J. G. Dai and P. Prabhakar. The throughput of data switches with and without speedup. *IEEE INFOCOM*, pages 556–564, 2000.
3. A. Stolyar. Maxweight scheduling in a generalized switch: State space collapse and equivalent workload minimization in heavy traffic. *Annals of Applied Probability*, 4(1):1–53, 2004.
4. K. Ross and N. Bambos. Local search scheduling algorithms for maximal throughput in packet switches. *IEEE INFOCOM*, 2004.
5. M. Karol, M. Hluchyj, and S. Morgan. Input vs. output queueing on a space-division packet switch. *IEEE Trans. Communications*, 25(12), 1987.
6. N. McKeown. The islip scheduling algorithm for input queued switches. *IEEE Trans. on Networking*, 7:188–201, 1999.
7. L. Tassiulas. Linear complexity algorithms for maximum throughput in radio networks and input queued switches. *IEEE INFOCOM*, pages 533–539, 1998.
8. K. Ross, N. Bambos, K. Kumaran, I. Suniee, and I. Widjaja. Dynamic scheduling of optical data bursts in time-domain wavelength interleaved networks. *IEEE Hot Interconnects*, pages 108–113, 2003.
9. K. Ross, N. Bambos, K. Kumaran, I. Suniee, and I. Widjaja. Scheduling bursts in time-domain wavelength interleaved networks. *IEEE Journal on Selected Areas in Communications*, 21(9):1441–1451, 2003.
10. I. Widjaja, I. Saniee, R. Giles., and D. D. Mitra. Light core and intelligent edge for a flexible, thin-layered and cost-effective optical transport network. *IEEE Communications Magazine*, 45(5):31–36, 2003.
11. J. von Neumann. A certain zero-sum two-person game equivalent to the optimal assignment problem. *Contributions to the Theory of Games*, 2:5–12, 1953.
12. K. Ross. *Dynamic Scheduling in Queueing Systems with Applications to Communication Networks*. PhD thesis, Stanford University, 2004.

4

Geometry of Packet Switching: Maximal Throughput Cone Scheduling Algorithms

Kevin Ross[1] and Nicholas Bambos[2]

[1] University of California Santa Cruz, kross@soe.usc.edu
[2] Stanford University, bambos@stanford.edu

In this chapter, we discuss the key ideas underlying some recent developments in packet switching. They concern algorithms for scheduling packets through switching fabrics, primarily to maximize throughput and support differentiated quality of service.

We first develop a model capturing the packet queueing and scheduling dynamics of the switch in a 'vectorized' framework, which allows for the rich geometry of the switching problem to emerge. We then present a class of algorithms that dynamically schedule packets for transfer through the switching fabric, based on which conic space the packet backlog vector resides in. Appropriate construction of the cones leads to maximum throughput. Cone algorithms subsume the well known 'maximum weight matching' algorithms for packet switching as a special case. We discuss various aspects of cone algorithms including robustness, scalability, throughput and quality of service support.

4.1 Introduction

Packet switching is a basic technology of the Internet core. Scheduling packets to be transferred from switch input ports to output ones, by dynamically selecting the input–output port connectivity pattern of the switching fabric, is a key function of packet switches. Packet scheduling algorithms have recently attracted much attention in high-performance/speed switching [1–6]. In this chapter, we discuss some key ideas and results related to recent developments in this field.

At an appropriate level of operational abstraction, we can view a packet switch as a connector between incoming and outgoing network links. Packets arrive at input ports and are routed to the desired output ports via the switch interconnection fabric. Arriving packets are buffered (in general) at input ports to handle transfer resource contention in the switch fabric.

The simplest 1-by-2 switch of Figure 4.1 has a single input line/port that can be connected to either of two output lines/ports. A more complicated 2-by-2 switch is shown in Figure 4.2 and a general N-by-N switch in Figure 4.3. It has N input

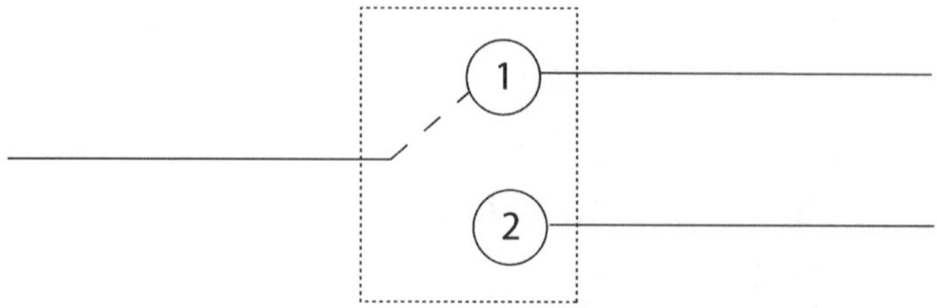

Figure 4.1. The simplest 1-by-2 switch, having a single input and two output lines. At each time slot, it can be configured to connect the incoming line to either of the two outgoing lines.

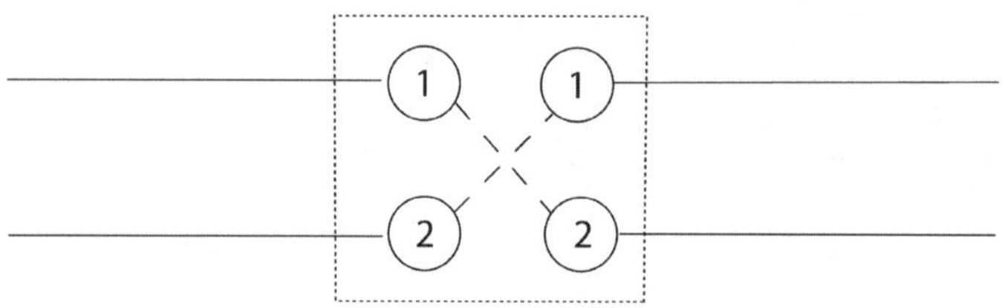

Figure 4.2. A simple 2-by-2 switch with two input and two output ports. At any time slot, each input port is connected to a single output port, either 1-to-1 and 2-to-2 or 1-to-2 and 2-to-1.

ports connecting to N output ports through the switch fabric, for example, a crossbar. Each output port connects to exactly one input port, and so there are various interconnection modes, corresponding to the input–output port matchings.

The decision of which output port each incoming packet is destined for is assumed to be predetermined in this discussion. This could be according to a predefined path the packet takes through a network, or independently made by a separate routing algorithm. We are interested in how to dynamically configure the connectivity pattern/mode of the switch given incoming packets desiring to reach given output ports.

There are several variations of the core switching abstraction discussed above, including multicast (replication of incoming packets at many output ports), multiclass (differentiating between various packet classes, *e.g.* voice, video, data, etc.), with speedup (the switch fabric is of higher rate than the lines), etc.

An important issue in switches is the potential head-of-line blocking. If packets destined for distinct output ports are queued up in the same input buffer and processed in a FIFO manner, then a packet at the head of the FIFO line can block service to other packets behind it, reducing the overall switch throughput [7]. This can be overcome by using *virtual output queues* (VOQ), as illustrated for a 2-by-2 switch in Figure 4.4. Instead of having one buffer for each input port, the input port

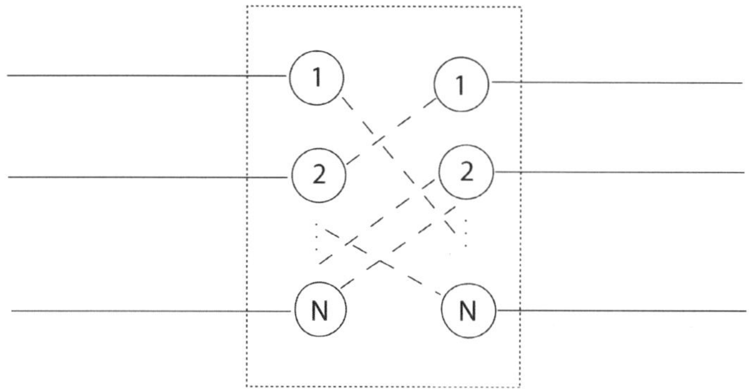

Figure 4.3. An N-by-N switch connecting each input port to exactly one output port. Since each port can have exactly one connection, the set of possible modes for the switch corresponds to the matchings of input and output ports.

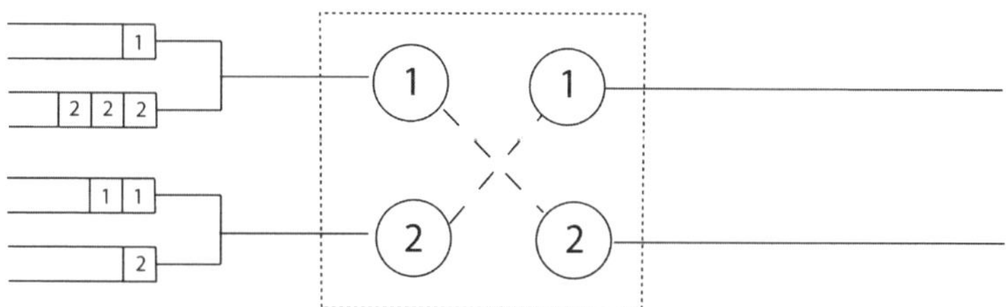

Figure 4.4. To avoid head-of-line blocking, each input–output pair has its own input queue or virtual output queue (VOQ)

buffer is separated into parallel queues, one for each output port. Thus there is a queue for each input–output pair.

The concept of virtual output queues can be generalized further. For general switches, we consider a *separate* queue for each *unique* traffic type. For example, if arriving packets belong to different classes, we use a separate queue per class. If some packets are to be multicast to some output ports, then they are queued up in a separate buffer. This generalized form of a virtual output queue avoids potential head-of-line blocking effects, since every packet in each one queue is of identical type.

Several important issues emerge when one considers the control of packet flows through switches, including packet scheduling to maximize switch throughput and provide quality of service to traffic flows. We focus on such issues in this chapter.

In Section 4.2 we develop a general switching model. The issues of switch capacity and throughput are discussed in Section 4.3. A key family of algorithms, called Projective Cone Scheduling (PCS) algorithms, is presented in Section 4.4. They max-

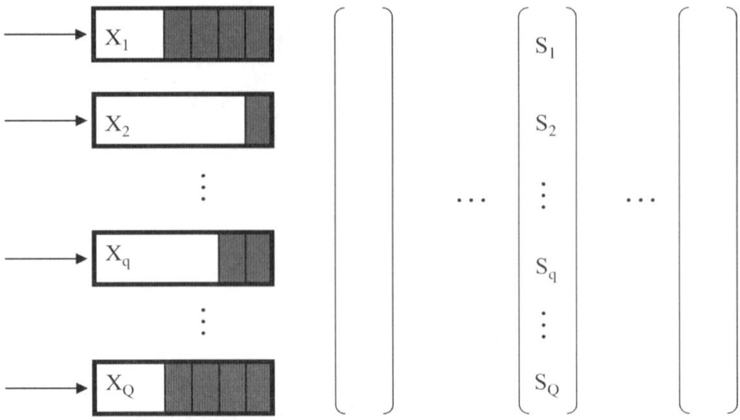

Figure 4.5. The general switch model is illustrated above. Arriving packets/cells are stored in (parallel) queues. There is a finite set of available service modes to choose from. Each mode corresponds to a combination of cells (S_q from each queue q) served/transferred in a time slot.

imize the switch throughput. A key extension, called delayed PCS, is also discussed. In Section 4.5, complexity and scalability issues are discussed and two other classes of scalable packet scheduling algorithms, called approximate PCS and local PCS correspondingly, are presented. They also maximize the switch throughput. The emphasis is on presenting the key ideas and using 'geometric' intuition as much as possible to illuminate the operational subtleties of the algorithms.

4.2 Backlog Dynamics of Packet Switches

We view the switch as a controlled queueing system. The general model is illustrated in Figure 4.5, having Q parallel queues and M switching/service modes to select from. The queues are labeled by $q \in \mathcal{Q} = \{1, 2, ..., Q\}$ and the modes by $m \in \mathcal{M} = \{1, 2, ...M\}$. Each queue buffers a type/class of packets. Time is divided into equal length time slots (arbitrarily small), labeled by $t = \{0, 1, 2, ...\}$. For simplicity, we assume that a data packet can be divided into workload units called cells (arbitrarily small). Each switching mode $m \in \mathcal{M}$ corresponds to a service vector

$$S^m = (S_1^m, S_2^m, ..., S_q^m, ..., S_Q^m), \tag{4.1}$$

where $S_q^m \geq 0$ is the number of cells that can be served/removed from queue q in a time slot, when the system operates under mode m. For example, for three queues $Q = 3$, the service mode $(1, 0, 5)$ refers to serving 1 cell from queue 1, 0 cells from 2, and 5 cells from 3, in a time slot when this mode is used. For standard crossbar packet switches, each input port may be connected to at most one output port and transfer one cell in a time slot, so the service vectors have only 0/1 entries (see Figure 4.6).

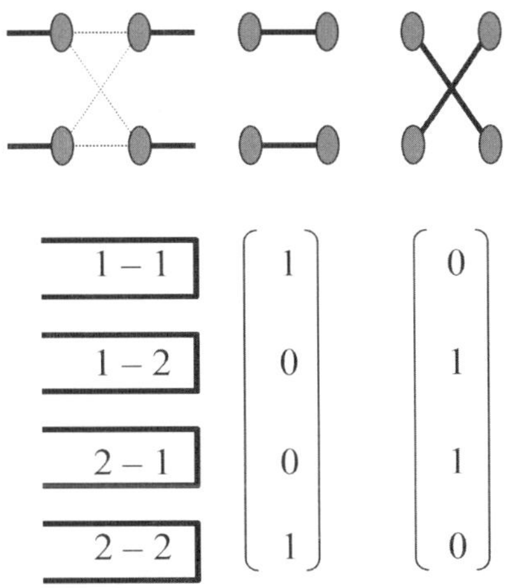

Figure 4.6. A simple example of a 2-by-2 switch and its two connectivity patterns (top row). There are 4 queues, one for each input–output port combination and two 4-dimensional 0/1 service vectors (lower row) corresponding to the connectivity patterns.

We assume that the set of service vectors is *complete*, that is, any projection sub-vector (with any entries set to zero) of a service vector is also an available service vector. Hence, when some queues become empty and naturally cease receiving service, the resulting effective service vector is a feasible one. In particular, the zero service is feasible. Note that, if an incomplete service set is given, we can naturally complete it by adding all projection sub-vectors of the initial ones.

Arriving cells are buffered immediately in their respective queues, and served in a first-in-first-out (FIFO) manner within each queue (actually, FIFO is not essential for the following results, but is used as a natural default queueing discipline). Let $A_q(t)$ be the number of cells arriving at queue q in time slot t. The *arrival vector* in time slot t is

$$A(t) = (A_1(t), A_2(t), ..., A_q(t), ..., A_Q(t)) \tag{4.2}$$

The *backlog vector* $X(t)$ of the queueing system at time t is

$$X(t) = (X_1(t), X_2(t), ..., X_q(t), ..., X_Q(t)) \tag{4.3}$$

where $X_q(t)$ is the number of cells (workload) waiting in queue $q \in \mathcal{Q}$ at time t. For any backlog vector X, the service modes allowable should be only those that could remove at most the number of cells residing at each queue, that is,

$$\mathcal{M}(X) = \{m \in \mathcal{M} : S^m \leq X\} \tag{4.4}$$

(where the vector inequality is considered componentwise). This makes sure that no allowable service vector will remove more cells than those available in a queue at each time slot.

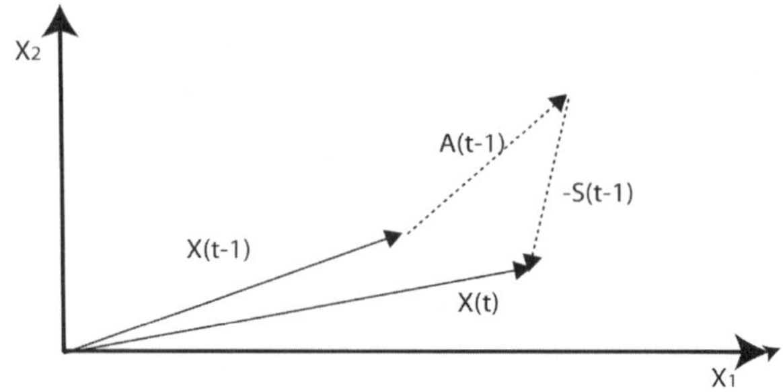

Figure 4.7. The evolution of the backlog $X(t)$ according to Equation (4.6). $A(t-1)$ is the arrival vector and $S(t-1)$ the service vector under our control.

We control the queueing system by choosing the service mode $m(t)$ of the system at time t, that is, its service vector

$$S(t) = (S_1(t), S_2(t), ..., S_q(t), ..., S_Q(t)) = S^{m(t)} \tag{4.5}$$

Given the mode/service control $S(t)$ the backlog state evolves (see Figure 4.7) according to the equation

$$\begin{aligned} X(t) &= X(t-1) + A(t-1) - S(t-1) \\ &= X(0) + \textstyle\sum_{s=0}^{t-1} A(s) - \sum_{s=0}^{t-1} S(s) \end{aligned} \tag{4.6}$$

for $t \in \{1, 2, ...\}$, where $X(0)$ is the initial backlog. The service is store-and-forward (although cut-through service can easily be accommodated by amending the evolution equation).

4.3 Switch Throughput and Rate Stability

To address the issue of what is the maximum load that the switch can handle, and under which mode control or cell scheduling policy, we first need to define precisely the concepts of switch load, stability and capacity.

We make the natural assumptions that the long-term average cell load for each $q \in \mathcal{Q}$ is well-defined, positive and finite, that is,

$$\lim_{t\to\infty} \frac{\sum_{s=0}^{t-1} A_q(s)}{t} = \rho_q \in (0, \infty) \tag{4.7}$$

and the load is bounded per slot or $A_q(t) \le A_q^{\max}$ for all t and some arbitrarily fixed positive ceiling $A_q^{\max}$ (for example, corresponding to the port line rate). The switch *load vector* is

$$\rho = (\rho_1, \rho_2, ..., \rho_q, ..., \rho_Q) \tag{4.8}$$

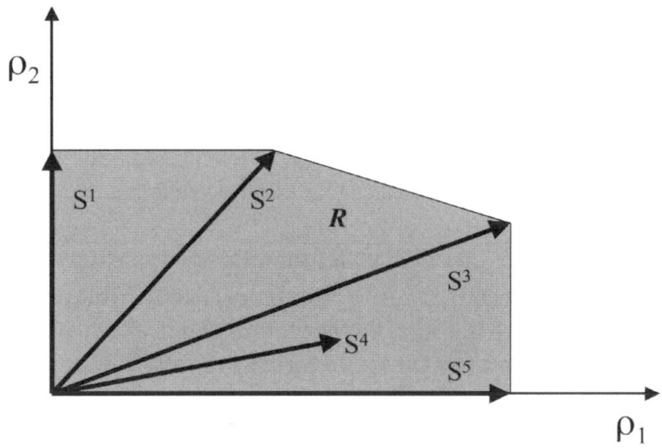

Figure 4.8. The stability region of a system of two queues and five service vectors $\{S^1, S^2, S^3, S^4, S^5\}$. If the load vector ρ is within the shaded region $\mathcal{R}$, then there exists a convex combination of service modes which would ensure rate stability by serving at least the arrival rate to each queue.

No further assumptions are imposed on traffic trace statistics, so traces to various queues can be highly interdependent indeed.

Note that the long-term average cell departure rate from each queue q is simply (when the limit exists)

$$\lim_{t\to\infty} \frac{\sum_{s=0}^{t-1} S_q(s)}{t} = \rho_q^{out} \tag{4.9}$$

when the service vector control or packet scheduling policy $S(t)$ is used. One desires and expects that under normal operation cell *flow conservation* is maintained through the switch, that is, the average cell departure rate equals the average arrival rate or

$$\rho_q^{out} = \lim_{t\to\infty} \frac{\sum_{s=0}^{t} S_q(s)}{t-1} = \lim_{t\to\infty} \frac{\sum_{s=0}^{t-1} A_q(s)}{t} = \rho_q \tag{4.10}$$

for each $q \in \mathcal{Q}$. In that case, we call the queueing system *rate stable*, in the sense that there is input–output flow equilibrium, as opposed to a flow deficit at the output. Note that rate stability (4.10) is equivalent to

$$\lim_{t\to\infty} \frac{X(t)}{t} = 0 \tag{4.11}$$

due to (4.6) and (4.7), that is, queue backlogs do not grow linearly over time.

The set of load vectors ρ for which the switch is rate stable (4.10) for some service control or scheduling policy $S(t)$ is known as the *stability region* $\mathcal{R}$. A rate vector is called allowable if it is within the stability region. It turns out that the stability region is

$$\mathcal{R} = \left\{ \rho : \rho \leq \sum_{m=1}^{M} \phi_m S^m \text{ for some } \phi_m \geq 0, \sum_{m=1}^{M} \phi_m = 1 \right\} \tag{4.12}$$

that is, $\rho \in \mathcal{R}$ if it is dominated by some convex combination of the service vectors (see Figure 4.8). Hence, if ρ known in advance, a scheduling policy which uses each S^m for a fraction ϕ_m of time will ensure that (4.10) is satisfied and the switch is rate stable.

It is interesting to note that if $\rho \notin \mathcal{R}$, then the switch is unstable[3] under *any* scheduling policy, that is, $\lim_{t\to\infty} \frac{X(t)}{t} \neq 0$ or (equivalently) $\limsup_{t\to\infty} \frac{X_q(t)}{t} > 0$ for some queue $q \in \mathcal{Q}$. The question then becomes under which 'smart' scheduling policies $S(t)$ the switch can remain stable for all $\rho \in \mathcal{R}$. This is the subject of the discussion in the next section.

Finally, it is interesting to note that we can characterize feasible load vectors $\rho \in \mathcal{R}$ in a 'geometric' manner. Specifically, if $\rho \in \mathcal{R}$, then for any Q-dimensional real vector[4] v, we have

$$\langle \rho, v \rangle \leq \max_{m\in\mathcal{M}} \langle S^m, v \rangle \tag{4.13}$$

Thus, in any direction v, the projection of a feasible load vector is dominated by the projection of some service vector. This is a property that proves useful later.

4.4 Cone Algorithms for Packet Scheduling

One intuitively expects a good scheduling algorithm to maximize the switch throughput by adapting to traffic fluctuations and handling excessive queue backlogs for the maximum possible mean traffic load, even if the latter is not *a priori* known or gradually shifts over time.

[3] A formal proof of that fact can be found in [6,8]. The following heuristic arguments, however, reveal the key intuition. Arguing by contradiction, suppose that $\lim_{t\to\infty} \frac{X(t)}{t} = 0$. Assuming we use a scheduling policy $S(t)$, from (4.6) we get $X(t) = X(0) + \sum_{s=0}^{t-1} A(s) - \sum_{s=0}^{t-1} \sum_{m=1}^{M} S^m \mathbf{1}_{\{S(s)=S^m\}}$. Dividing the inequality terms through by t, taking the limit as $t\to\infty$, using (4.7) and rearranging the terms, we get $\rho \leq \sum_{m=1}^{M} S^m \lim_{t\to\infty} \frac{1}{t} \sum_{s=0}^{t-1} \mathbf{1}_{\{S(s)=S^m\}}$ Setting $\phi_m = \lim_{t\to\infty} \frac{1}{t} \sum_{s=0}^{t-1} \mathbf{1}_{\{S(s)=S^m\}}$, it follows that $\rho \leq \sum_{m=1}^{M} S^m \phi_m$ with $\sum_{m=1}^{M} \phi_m = 1$, which contradicts the assertion that $\rho \notin \mathcal{R}$ according to (4.12). We have 'loosely' assumed in the heuristic arguments above that the limits $\phi_m = \lim_{t\to\infty} \frac{1}{t} \sum_{s=0}^{t-1} \mathbf{1}_{\{S(s)=S^m\}}$ exist. This may not be the case and so one has to work with subsequences on which these limits do exist in the formal proof.

[4] A formal proof of this property can be found in [5,8,9]. A brief sketch of the proof, however, is given below. If $\rho \in \mathcal{R}$, then $\rho \leq \sum_{m=1}^{M} \phi_m S^m$; hence, $\rho_q \leq \sum_{m=1}^{M} \phi_m S_q^m$ for each queue q. Consider now any arbitrarily fixed real vector $v = (v_1, v_2, .., v_q, ..., v_Q)$. We have $\rho_q v_q \leq \sum_{m=1}^{M} \phi_m S_q^m \mathbf{1}_{\{v_q \geq 0\}} v_q$ for each queue, since $\rho_q \geq 0$ and $S_q \geq 0$. Writing $V(S) = (S_q \mathbf{1}_{\{v_q\geq 0\}},\ q \in \mathcal{Q})$, we see that $\langle \rho, v \rangle \leq \sum_{m=1}^{M} \phi_m \langle V(S^m), v \rangle \leq \max_{m\in\mathcal{M}} \langle S^m, v \rangle$, where the right inequality is due to the assumed completeness of the service vector set.

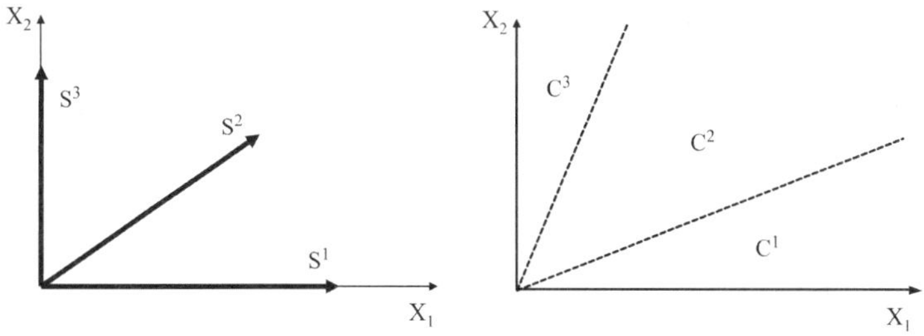

Figure 4.9. The cone geometry of the PCS algorithm. In this example, the system has two queues and three service vectors $S^1 = (4, 0), S^2 = (3, 2), S^3 = (0, 3)$ (and their projection sub-vectors for completion). Note that S^1 focuses on queue 1 and S^3 on queue 2, while S^2 serves both in a more balanced manner. The first graph shows the service vectors available. The second graph shows the service cones $\mathcal{C}^m$ when the identity matrix $\mathbf{B} = \mathbf{I}$ is used. For example, if the workload X is within the cone $\mathcal{C}^1$ then service mode S^1 is used. Whenever the workload drifts into a new cone, the new service mode is selected.

4.4.1 Projective Cone Scheduling (PCS)

We now discuss a rich family of high-performance scheduling algorithms, which are called *Projective Cone Scheduling (PCS)* algorithms. They are *projective* because at any time they select the service vector S^m with maximum projection $\langle S^m, \mathbf{B}X\rangle$ onto the (twisted) backlog vector $\mathbf{B}X$, for a some fixed twisting $Q \times Q$ matrix $\mathbf{B}$. Specifically, when the backlog is X, the PCS algorithm selects and activates a service mode m^* such that

$$\left\langle S^{m^*}, \mathbf{B}X \right\rangle = \max_{m \in \mathcal{M}(X)} \langle S^m, \mathbf{B}X\rangle \tag{4.14}$$

given a fixed matrix $\mathbf{B}$ (with properties discussed below). That is, when the backlog is X, the PCS algorithm selects a mode in the set

$$\mathcal{M}^*(X) = \left\{ m^* \in \mathcal{M}(X) : \left\langle S^{m^*}, \mathbf{B}X \right\rangle = \max_{m \in \mathcal{M}(X)} \langle S^m, \mathbf{B}X\rangle \right\} \tag{4.15}$$

of all modes whose service vectors have maximal projection on $\mathbf{B}X$. There may be more than one such mode, in which case any one is selected arbitrarily.

The family of PCS algorithms has a nice geometric interpretation (see Figure 4.9) which justifies the name *cone algorithms*. For each service mode $m \in \mathcal{M}$, let $\mathcal{C}^m$ be the set of backlog vectors X for which m would be chosen under the PCS algorithm; that is,

$$\mathcal{C}^m = \{X : m \in \mathcal{M}^*(X)\} \tag{4.16}$$

This is the set of backlogs X for which $\langle S^m, \mathbf{B}X\rangle$ is maximized for S^m or the set of backlogs for which the PCS algorithm chooses the service vector S^m. Note that $\mathcal{C}^m$ is a geometric cone. Indeed, if $X \in \mathcal{C}^m$ then $\alpha X \in \mathcal{C}^m$ for any positive scalar α

(since $\langle S^m, \mathbf{B}X \rangle$ being maximal implies that $\langle S^m, \mathbf{B}\alpha X \rangle$ is maximal). In view of the above, we can now define the PCS algorithm as

$$\text{PCS}: \text{ when } X \in \mathcal{C}^m \tag{4.17}$$

The cones $\{\mathcal{C}^m, m \in \mathcal{M}\}$ form a partition of the backlog space and the system backlog evolves by drifting in and across these cones as packets arrive and leave after being served. Illustrative examples of the PCS cone geometry are shown on Figures 4.10 and 4.11.

Note that PCS algorithms require *no knowledge of the load vector* ρ. They directly select the service mode m based on the current workload X, depending on which cone $\mathcal{C}^m$ the latter belongs to. By selecting and fixing $\mathbf{B}$ we can generate a very rich family of PCS algorithms. Which of those, however, achieve maximal throughput? We address this issue below.

It turns out that if $\mathbf{B}$ is (1) *positive definite*, (2) *symmetric* and (3) has *negative or zero off-diagonal elements*, then the corresponding PCS algorithm maximizes the switch throughput; that is,

$$\lim_{t \to \infty} \frac{X(t)}{t} = 0, \text{ for all } \rho \in \mathcal{R} \tag{4.18}$$

or rate stability and flow conservation is maintained for all traffic loads in $\mathcal{R}$. The identified properties of $\mathbf{B}$ imply that the diagonal elements of $\mathbf{B}$ must be positive. Such matrices are known as *Stieltjes* matrices [10]. The formal proof of (4.18) can be found in [11], but a heuristic one revealing the key intuition is provided below.

Let us consider some special – yet interesting – cases, emerging when $\mathbf{B} = \mathbf{I}$ (the identity matrix). First, when the switch is Q-by-1 (having Q parallel queues and a single server that can be allocated to any queue to provide service at rate 1), then PCS is equivalent to the Largest-Queue-First scheduling discipline. That is, PCS allocates the server to the queue with the largest current backlog. Second, when the switch is a crossbar Q-by-Q one (with 0/1 service vectors), then PCS with $\mathbf{B} = \mathbf{I}$ is equivalent to the well known Maximum-Weight-Matching (MWM) algorithm [1].

4.4.2 Relaxations, Generalizations, and Delayed PCS (D-PCS)

There is an interesting relaxation of the PCS algorithm, which allows for generalizing the throughput maximization result to a far broader class of systems, reflecting the high robustness of the core PCS algorithm. This is called *Delayed PCS (D-PCS)* and is explained below.

Recall first that the basic PCS scheme selects mode m immediately when $X \in \mathcal{C}^m$. To define D-PCS, we relax this requirement, by allowing an arbitrarily fixed time-lag K in selecting a mode m after the backlog X enters cone $\mathcal{C}^m$ and drifts in it. That is, if $X(t)$ enters cone $\mathcal{C}^m$ at some time t_o and stays drifting in it for more than D time slots, then D-PCS is guaranteed to select mode m after at most K slots and keep using it as long as $X(t)$ stays in $\mathcal{C}^m$. Formally,

$$\text{if } X(t) \in \mathcal{C}^m \text{ for } t \in \{t_o, t_o{+}1, ..., t_o{+}K, ...\}, \text{ then use } m \text{ for } t \geq t_o{+}K \tag{4.19}$$

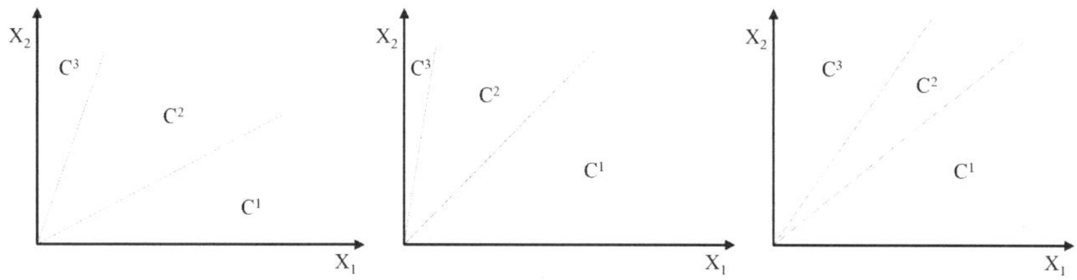

Figure 4.10. The PCS cone geometry for a system of two queues and three service vectors $S^1 = (4,0), S^2 = (3,2), S^3 = (0,3)$. Note that S^1 serves queue 1 and S^3 queue 2 only, while S^2 serves both in a somewhat balanced manner. The graphs show the cones $\mathcal{C}^m$ when each of the matrices $\mathbf{B}_1 = \begin{bmatrix} 1 & 0 \\ 0 & 1 \end{bmatrix}$, $\mathbf{B}_2 = \begin{bmatrix} 2 & 0 \\ 0 & 1 \end{bmatrix}$ and $\mathbf{B}_3 = \begin{bmatrix} 1 & -0.5 \\ -0.5 & 1 \end{bmatrix}$ is used. For example, if $X \in \mathcal{C}^1$ then service mode S^1 is selected by PCS. Changing the top diagonal element from 1 to 2 results in shrinking cone $\mathcal{C}^3$ and expanding cone $\mathcal{C}^1$, giving higher 'priority' to queue 1 and using S^1 over a larger space of workloads. In the third plot, where negative off-diagonal elements are introduced, the middle cone $\mathcal{C}^2$ shrinks. This corresponds to 'coupling' the two queues tighter and pulling the two workloads closer together by trying to load-balance them.

The time-lag/delay D in selecting the right mode is fixed, but arbitrary; varying it generates a family of D-PCS algorithms.

In summary, D-PCS will track the mode choices of PCS with a time-lag, selecting the same (right) mode when the backlog has remained within a single cone for at least K time slots, for some fixed K. The modes chosen by D-PCS in the first K time slots after entering a cone can be arbitrary

This relaxation allows us to consider a number of important generalizations to the switch structure. These include: (1) the backlog information provide to the scheduler is outdated and delayed; (2) packets cannot be divided into individual cells, but have to be processed non-preemptively; (3) the scheduling decisions (computation of m) can only be done periodically and not at every time slot; (4) only subsets (neighborhoods) of service modes, rather than the whole set $\mathcal{M}(X)$, can be used at each time (see local-PCS later), etc. All the above generalizations allow for applying the D-PCS algorithm, tracking the PCS mode choices with a time lag, caused by a delay in information gathering or due to limited choices at each time slot.

As with PCS, it turns out that if $\mathbf{B}$ is (1) *positive definite*, (2) *symmetric* and (3) has *negative or zero off-diagonal elements*, then the corresponding D-PCS algorithm maximizes the switch throughput; that is,

$$\lim_{t\to\infty} \frac{X(t)}{t} = 0, \text{ for all} \tag{4.20}$$

$$\rho \in \mathcal{R} \tag{4.21}$$

or rate stability and flow conservation is maintained for all traffic loads in $\mathcal{R}$. The formal proof of (4.20) can be found in [11], but a heuristic sketch of it is provided below.

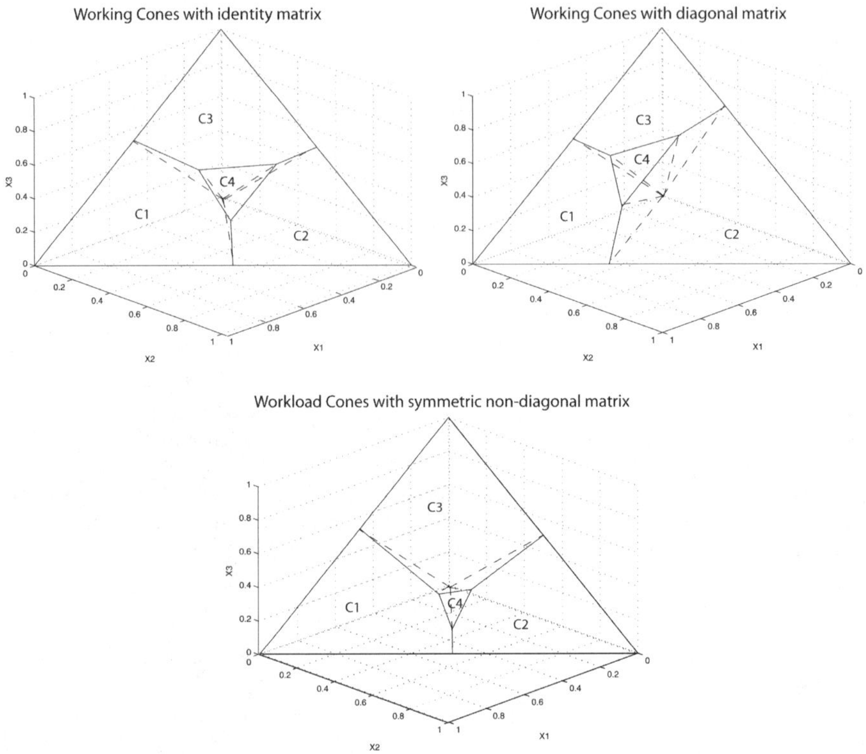

Figure 4.11. The PCS cone geometry for a system of three queues and four service vectors $S^1 = (9,0,0)$, $S^2 = (0,8,0)$, $S^3 = (0,0,8)$, $S^4 = (3,4,3)$. The cones are visualized by considering their intersection locus with the plane defined by the points $(1,0,0), (0,1,0), (0,0,1)$ or by the equation $(1,1,1)X = 1$. The graphs show the $\mathcal{C}^m$ service cones when the matrices $\mathbf{B}_1 = \begin{bmatrix} 1 & 0 & 0 \\ 0 & 1 & 0 \\ 0 & 0 & 1 \end{bmatrix}$, $\mathbf{B}_2 = \begin{bmatrix} 1 & 0 & 0 \\ 0 & 2 & 0 \\ 0 & 0 & 1 \end{bmatrix}$, and $\mathbf{B}_3 = \begin{bmatrix} 1 & -0.5 & 0 \\ -0.5 & 1 & 0 \\ 0 & 0 & 1 \end{bmatrix}$ are used correspondingly. For example, if $X \in \mathcal{C}^1$ then S^1 is used. In the second figure, the weight of the second queue being 2 (compared to 1 for the others), increases the service priority of queue 2 and enlarges the cone $\mathcal{C}^2$ by pushing away its boundaries to neighboring cones (and squeezing the latter). The third plot has negative off-diagonal elements causing X_1 and X_2 to counter each other and become more tightly coupled or entangled.

4.4.3 Argument Why PCS and D-PCS Maximize Throughput

It is interesting to see why PCS and D-PCS maximize the switch throughput, when the fixed matrix $\mathbf{B}$ is positive definite, symmetric, and has negative or zero off-diagonal elements, by forcing

$$\lim_{t\to\infty} \frac{X(t)}{t} = 0 \tag{4.22}$$

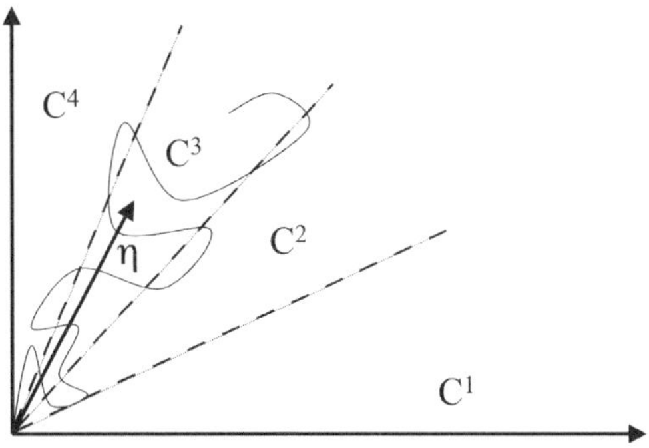

Figure 4.12. The backlog drifting in and across the mode cones partitioning the backlog space, eventually 'explodes' along the direction η, as considered in Section 4.4.3

for all $\rho \in \mathcal{R}$. We highlight the key intuition below, using heuristic arguments to provide a sketch of the proof which can be found in [5, 9, 11].

Arguing by contradiction, suppose (loosely) that $\lim_{t\to\infty} \frac{X(t)}{t} = \eta > 0$ (componentwise) for some $\rho \in \mathcal{R}$. Let η be in the cone C^{m^*} where the mode m^* and service vector S^{m^*} are used by PCS and (eventually) D-PCS (see Figure 4.12). Therefore, there is a finite time T after which both algorithms use S^{m^*} consistently. Hence, $X(t) = X(T) + \sum_{s=T}^{t-1}(A(s) - S(s)) = \sum_{s=T}^{t-1} A(s) - (t-T)S^{m^*}$. Dividing by t and letting $t\to\infty$, we get $\lim_{t\to\infty} \frac{X(t)}{t} = \rho - S^{m^*}$ or $\eta = \rho - S^{m^*}$. Projecting on $\mathbf{B}\eta$, we get

$$\langle \eta, \mathbf{B}\eta \rangle = \langle \rho, \mathbf{B}\eta \rangle - \left\langle S^{m^*}, \mathbf{B}\eta \right\rangle = \langle \rho, \mathbf{B}\eta \rangle - \max_{m\in\mathcal{M}} \langle S^m, \mathbf{B}\eta \rangle \tag{4.23}$$

recalling that S^{m^*} should have maximal projection on $\mathbf{B}\eta$ since $\eta \in C^{m^*}$. From (4.23), setting $v = \mathbf{B}\eta$, the right-hand side must be negative or zero, hence, $\langle \eta, \mathbf{B}\eta \rangle \leq 0$. This implies that $\eta = 0$, since $\mathbf{B}$ is positive definite, contradicting the assertion that $\eta > 0$ and establishing contradiction.

The formal proof is a bit more subtle. Indeed, one has to consider the convergence $\frac{X(t)}{t}\to\eta$ on convergent subsequences, since the direct limit may not exist. Moreover, some components of η may be 0, although $\eta \neq 0$ overall. To clinch the full proof one must then invoke the three aforementioned properties of $\mathbf{B}$.

Finally, it is interesting to note that the service vectors drain the backlog by attracting it towards the cone boundaries. This is clearly demonstrated in the example shown in Figure 4.13.

4.4.4 Quality of Service and Load Balancing

The choice of the fixed matrix $\mathbf{B}$ in the PCS and D-PCS algorithms has implications for quality of service received by each queue and load balancing across various

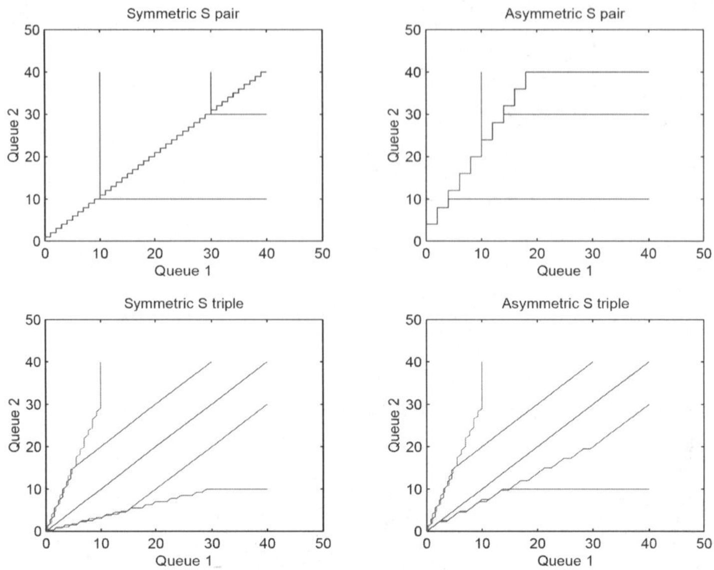

Figure 4.13. Backlog draining under PCS for a simple example of two queues, which drain from different initial backlogs, while no arrivals are allowed. The identity matrix $\mathbf{B} = \mathbf{I}$ is used and the set of service vectors is varied over the four plots. Each plot shows how the two queues drain under PCS from five different initial workload levels. The upper-left plot has two service vectors $S^1 = (1, 0)$ and $S^2 = (0, 1)$ available. The upper-right plot has $S^1 = (2, 0)$ and $S^2 = (0, 1)$. The lower-left plot has three service vectors to select from, $S^1 = (1, 0), S^2 = (0, 1)$ and $S^3 = (0.75, 0.75)$. The lower-right plot has the vectors $S^1 = (1.25, 0), S^2 = (0, 1)$ and $S^3 = (0.75, 0.75)$. As the set of service vectors changes, the drain trajectories change too. They are always attracted, however, towards the cone boundaries and follow them after hitting them initially.

queues. The matrix introduces several design degrees of freedom, providing Q^2 tunable parameters.

We can view $\mathbf{B} = \{B_{pq}\}$ in terms of the transformation it applies to the cone space, as illustrated in Figures 4.9, 4.10 and 4.11. When $\mathbf{B}_{qq}$ is increased (while other elements remain constant), queue q gains service priority and sees a lower average backlog. On the other hand, when $\mathbf{B}_{pq} < 0$ with $p \neq q$, the relative priority of queue p decreases, as the workload of queue q increases. Service attention shifts toward queue q, as X_q grows and away as X_p grows.

Overall, PCS dynamically adapts to backlogs and load balances across the various queues. Indeed, as the backlog of a queue increases excessively, PCS shifts attention to it and selects a service vector that drains it faster, potentially at the expense of other less congested queues which drain slower.

4.5 Complexity in Cone Schedules – Scalable PCS Algorithms

The computational complexity of finding the maximum $\langle S^m, \mathbf{B}X \rangle$ value over $m \in \mathcal{M}(X)$ in real-time (in each time slot) can be considerable. There is typically a huge number of service modes in $\mathcal{M}(X)$, which makes the realtime computation of the maximum prohibitive. For example, a switch with N input and N output ports ($Q = N^2$ virtual output queues) has service vectors corresponding to input–output port matchings, whose number grows factorially in N.

We outline two families of throughput-maximizing PCS relaxations to counter the challenge of complexity. These are known as *approximate PCS* and *local PCS* algorithms. The former reduce the regularity or accuracy with which $\langle S^m, \mathbf{B}X \rangle$ needs to be calculated and maximized, while the latter reduce the mode set over which maximization is considered at each time slot [5]. They are both *delayed PCS* algorithms (see Section 4.4.2) and, hence, achieve maximal throughput.

4.5.1 Approximate PCS

In Section 4.4.2 it was noted that the maximization of $\max_{m \in \mathcal{M}(X)} \langle S^m, \mathbf{B}X \rangle$ need not be under exactly up to date information. One natural option could be for the maximum to only be calculated every K with outdate backlog information, leading to a D-PCS algorithm.

A nice alternative is to choose at every time slot a near-optimal service vector S^{m^o} in the sense that

$$\left\langle S^{m^*}, \mathbf{B}X \right\rangle - E \leq \langle S^m, \mathbf{B}X \rangle \leq \left\langle S^{m^*}, \mathbf{B}X \right\rangle \tag{4.24}$$

where S^{m^*} is the actual one of maximal projection on $\mathbf{B}X$ and E is an arbitrarily fixed error term or slack. This near-optimal may be computed significantly faster than the optimal one and we can establish maximal throughput for this approximate PCS too, via a D-PCS type argument [8].

Figure 4.14 shows the performance of a queuing system under two different approximations described above, as well as the standard PCS. As expected, the latter outperforms both approximate PCS algorithms in terms of backlog; all three, however, can achieve maximal throughput.

4.5.2 Local PCS

In addition to the frequency of calculations, the number of modes under consideration at each time slot increases the complexity. The PCS cone geometry leads to intuition which helps alleviate this complexity. It is presented in Figure 4.15 for a simple two-queue example.

Suppose that at time $t-1$ the backlog $X(t-1)$ is in the cone $\mathcal{C}^m$. During time slot $t-1$, packets arrive to the various system queues, which cause the workload at the start of time t to jump to $X' = X - S + A(t-1)$ which belongs to the cone $\mathcal{C}^{m'}$ and the service vector used at X' is S'. Given assumed upper bounds A_q^{max} of

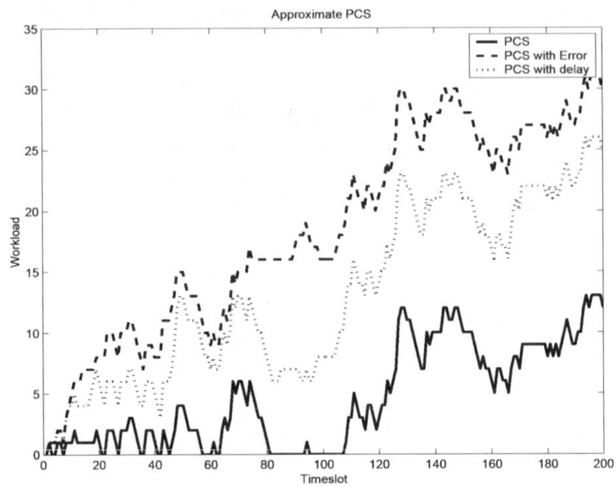

Figure 4.14. The same arrival trace applied to the same system of three queues under (1) standard PCS (solid line), (2) approximate PCS with delay of 10 slots (dotted line), and (3) approximate PCS with error/slack of 10 (dashed line). There are three queues and three service modes (each serving one queue). The total backlog over all three queues is shown in the figure and $\mathbf{B} = \mathbf{I}$.

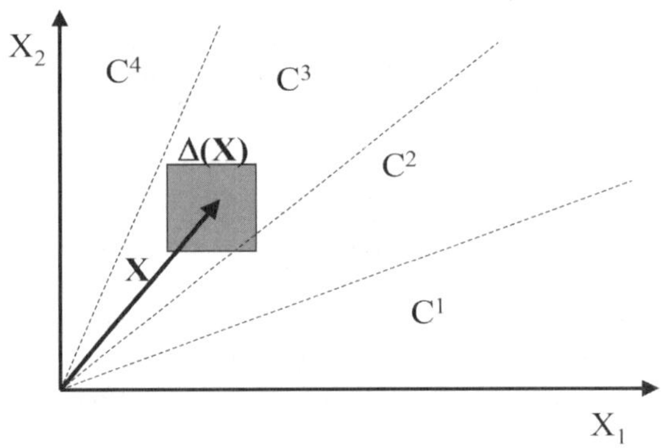

Figure 4.15. This figure shows the PCS cone structure for a simple example of two queues and four service cones. The jump in the backlog X in a slot is within the fixed displacement box $\Delta(X)$ of possible extreme values for $(A-S)$. As the workload X increases, the displacement box $\Delta(X)$ eventually intersects only cones that are neighbors to that where X resides.

arrivals per slot and the maximum possible departures per slot, the backlog X' has to be in a fixed 'displacement box' $\Delta(X)$ around X. Observe now the following:

1. If $|X|$ is small, then X' is in $\Delta(X)$ which intersects lots of cones $\mathcal{C}^{m'}$ that are reachable from X.
2. As $|X|$ gets larger, then X' is in $\Delta(X)$ which intersects fewer and fewer cones. Eventually, for large enough X, $\Delta(X)$ will intersect at most *neighboring* cones of $\mathcal{C}^m$. Thus, it suffices to compute $\langle S^m, \mathbf{B}X' \rangle$ and obtain the maximum over m' for which $\mathcal{C}^{m'}$ is adjacent to $\mathcal{C}^m$. This reduces computational complexity substantially.

This provides the key intuition for designing local PCS algorithms, because the concept of *adjacent cones* maps directly to that of local search in the space of service modes $\mathcal{M}$, as articulated below.

Define two service combinations S^m and $S^{m'}$ to be neighboring or adjacent if their corresponding cones $\mathcal{C}^m$ and $\mathcal{C}^{m'}$ are adjacent, in the sense that they share a non-trivial $(Q-1)$-dimensional common boundary. Let $\mathcal{N}_m$ be the set of modes whose corresponding service vectors are neighbors to S^m.

Define now the *Local PCS* algorithm inductively, as follows: if the service mode m was used in the previous time slot and the backlog is X' in the current time slot, then select a service mode m' in the current time slot such that

$$\langle S^m, \mathbf{B}X' \rangle = \max_{n \in \mathcal{N}_m} \langle S^m, \mathbf{B}X' \rangle \tag{4.25}$$

for a fixed matrix $\mathbf{B}$. This is quite similar to the definition of the standard PCS, *except* that the maximization is done *only* over the neighbors $\mathcal{N}_m$ of the previously used service mode m, as opposed to all modes in $\mathcal{M}(X)$ that would be standard. The set $\mathcal{N}_m$ typically has far fewer modes than $\mathcal{M}(X)$, hence the substantial reduction in complexity.

There is an interesting graph representation of the service mode set $\mathcal{M}$, which allows us to visualize the local PCS algorithm, as illustrated in Figure 4.16. Each service mode m corresponds to a distinct node of this graph, and an edge between two nodes m and m' exists if they are neighboring modes (*i.e.* $\mathcal{C}^m$ and $\mathcal{C}^{m'}$ are adjacent cones). Given that the local PCS algorithm is currently using mode m, it only needs to compute $\langle S^m, \mathbf{B}X' \rangle$ on the neighbors m' of node m on this graph and move to the node where this inner product is maximized. The neighbors m' could be stored in a lookup table or some graph data structure.

The local PCS algorithm, as described in (4.25), admits several extensions outlined below. (1) Instead of searching immediate neighbors of the previous mode, consider up to kth-degree neighbors (for example, 2nd-degree neighbors are separated by two edges in the graph shown in Figure 4.16). (2) Search neighbors until one is found which just improves $\langle S, \mathbf{B}X' \rangle$ rather than finding the maximizing neighbor. (3) Consider just one (or generally k) new mode(s) in each time slot and use it if it provides improvement, in such a way that every mode is considered within some bounded number of time slots. For example, do round-robin $t(\mod M)+1$ for an ordering $\{S^m\}_{m=1}^M$ of the service modes.

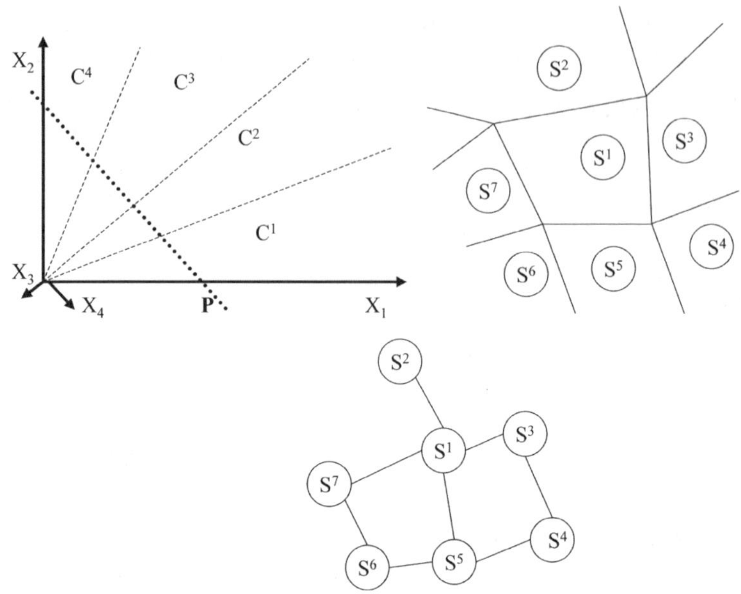

Figure 4.16. The local PCS algorithm can be visualized graphically. The first figure shows the PCS cone structure. The second figure shows the projection of the workload space onto a plane (such as P in the first graph). In the third figure, the mode graph is shown. Each node corresponds to a unique service mode m (or service vector S^m). Two nodes are joined by an edge if the cones corresponding to their service modes/vectors are adjacent. Given that the local PCS algorithm is currently using mode m, it only needs to compute $< S^{m'}, \mathbf{B}X'$ on the neighbors m' of node m on this graph and move to the node where this inner product is maximized.

All the previous algorithms, both approximate and local PCS versions, are actually delayed PCS algorithms. Therefore, if the matrix $\mathbf{B}$ is (1) positive definite, (2) symmetric, and (3) has negative or zero off-diagonal elements, then each of these algorithms will maximize the throughput of the system. The more 'relaxed' an algorithm is, the lower its implementation complexity will be, but also the worse it is expected to perform in terms of average queue backlogs.

4.6 Final Remarks

We have discussed some recent developments in packet switching algorithms having maximal throughput. Based on some simple ideas, we have developed a suite of algorithms, starting from the basic Projective Cone Scheduling one and considering a number of key relaxations that do not compromise the throughput. They lower, however, the implementation complexity, possibly at the expense of higher average backlogs.

How should one choose the matrix $\mathbf{B}$, so as to affect the quality of service received by each queue? Despite the fact we have demonstrated that this is doable, we have not provided a systematic way of selecting the matrix. This is the object of our current research on switching systems. Several challenges remain to be addressed in this direction in order to be able to systematically design the matrix $\mathbf{B}$ to satisfy a given set of quality of service constraints.

References

1. N. McKeown, A. Mekkittikul, V. Anantharam, and J. Walrand. Achieving 100% throughput in an input-queued switch. *IEEE Transactions on Communications*, 47(8):1260–1267, 1999.
2. L. Tassiulas. Linear complexity algorithms for maximum throughput in radio networks and input queued switches. *IEEE INFOCOM*, 533–539, 1998.
3. L. Tassiulas and A. Ephremides. Stability properties of constrained queueing systems and scheduling policies for maximum throughput in multihop radio networks. *IEEE Transactions on Automatic Control*, 37(12):1936–1948, 1992.
4. K. Ross and N. Bambos. Projective cone schedules in queueing structures; geometry of packet scheduling in communication network switches. *Allerton Conference on Communication, Control and Computing*, pages 626–635, 2002.
5. K. Ross and N. Bambos. Local search scheduling algorithms for maximal throughput in packet switches. *IEEE INFOCOM*, 1158-1169, 2004.
6. M. Armony and N. Bambos. Queueing dynamics and maximal throughput scheduling in switched processing systems. *Queueing Systems*, 44(3):209, 2003.
7. M. Karol, M. Hluchyj, and S. Morgan. Input vs. output queueing on a space-division packet switch. *IEEE Trans. Communications*, 25(12), 1347-1356, 1987.
8. K. Ross and N. Bambos. Projective processing schedules in queueing structures; applications to packet scheduling in communication network switches. Technical Report NETLAB-2002-05/01, Stanford University Engineering Library, 2002.
9. K. Ross and N. Bambos. Quality of service in packet switch scheduling. *IEEE ICC*, 1986-1990, 2004.
10. T.J. Stieltjes. Sur les racines de l'equation $x_n = 0$. *Acta Math.*, 9:385–400, 1887.
11. K. Ross. *Dynamic Scheduling in Queueing Systems with Applications to Communication Networks*. PhD thesis, Stanford University, 2004.

5

Fabric on a Chip: A Memory-management Perspective

Itamar Elhanany[1], Vahid Tabatabaee[2], and Brad Matthews[1]

[1] Department of Electrical and Computer Engineering, The University of Tennessee, 401-A Ferris Hall, Knoxville, TN 37996-2100. {itamar,bradmatthews}@ieee.org

[2] Institute for Advanced Computer Studies, University of Maryland at College Park, College Park, MD 20742, vahid@eng.umd.edu

This chapter introduces the background and motivation for, as well as the engineering challenges related to the notion of implementing a switch fabric on silicon. A review of prior work supporting the concept of fabric on a chip is provided leading to a consideration of the core challenge of designing scalable, high-performance memory management algorithms that can support next-generation services while facilitating the consolidation of various switching functions on a single chip. The methodology presented is illustrated using several practical as well as academic examples.

5.1 Introduction

Recent years have witnessed unprecedented advances in the design, verification formalism [1], and deployment of high-capacity, high-performance packet-switching fabrics. Such fabrics are commonly employed as the fundamental building blocks in data networking platforms that span a wide variety of application spaces. The market segment for which a product was designed predominantly governs the capacity of modern routers and switches. Core (or backbone) Internet routers, for example, are able to support multiple terabits per second [2], while systems built for metropolitan area networks (MANs) typically carry hundreds of gigabits per second (Gbps). Local area networks, representing the lower end of the switch/router market, have a switching fabric that supports up to tens of Gbps. However, switching fabrics are not limited to Internet transport equipment. Storage area networks (SANs), for example, often necessitate large packet switching engines to allow vast amounts of data to traverse a fabric, whereby storage data segments (*i.e.* blocks) flow from storage devices to servers and users, and vice versa.

During the late 1990s, many believed that the growth in Internet traffic would increase at a rate that would require significant upgrades in switching infrastructure as often as every 18 months. However, despite the increasing growth in user traffic (approximated to double every 12 months), the pragmatic requirements of backbone switches and routers are somewhat more modest. Nevertheless, the large number of components in switch fabrics, which drive such large systems, renders the latter highly complex to design, test and maintain. Thus, in an effort to alleviate some of the key difficulties in designing large switching fabrics, the concept of Fabric on a Chip (FoC) is introduced. In view of realistic technological limitations, it should be noted that FoC solutions would not be designed for core/backbone routers. Rather, the target application space for such products would be where hundreds of Gbps and below are required, *e.g.* in MAN, high-end LAN, and SAN, among others.

Taking advantage of recent advances in integrated circuit technologies, the goal of FoC architectures is to enable the consolidation of as many core switching functions as possible on a single chip. By achieving a high level of integration, it is argued that much larger systems can be readily realized. Moreover, the resulting designs will consume significantly fewer resources than the traditional approach. Accordingly, this chapter focuses on the topic of on-chip output-queued switch emulation. The memory-management problem is introduced, to which solutions using a novel architecture and algorithms are offered and discussed in detail.

5.1.1 Benefits of the Fabric-on-a-Chip Approach

There are numerous benefits to considering the notion of consolidating switching fabric functions on a chip. The first is the ability to reduce system physical components. By reducing the number of chips in the system, we directly obtain a reduction in size and design complexity, resulting in simplified board layouts and mechanical considerations. Such improvements in the design process are far from negligible, because a corollary to simplified design is shorter design cycles. As discussed earlier, reduction of power requirements – a crucial pragmatic aspect of any switch/router – is obtained due, primarily, to the reduction in high-speed serial transceivers in use. The proposed approach replaces chip-to-chip communication with on-chip communication, which has considerably lower power-consumption characteristics.

When considering the design of systems with capacities of hundreds of Gbps and beyond, other engineering aspects play a key role in guaranteeing successful deployment. One such key issue is staying within a workable power budget; high-speed serial transceivers that enable the transmission and reception of data signals at rates of Gbps, require a great deal of power. Practical maintenance constraints limit the amount of power that switches and routers may consume. In conventional switch/router designs, multiple high-bandwidth data signals originating from the input ports, arrive at the fabric and traverse it en route to their destination ports. Any reduction in the number of serializer/de-serializer (SerDes) circuits utilized by the various chips is guaranteed to directly reduce the overall system power consumption. Yet another element impacting power consumption is the amount of memory devices used.

A related advantage of FoC is higher reliability. It is generally acknowledged that lowering the number of (independent) components in any given system increases its reliability, since fewer components are prone to failures and thus need to be replaced. In view of recent technical standardization efforts pertaining to packet processing products, one may argue that FoC helps facilitate the rapid exploitation of standard interfaces to further support the interoperability between different semiconductor

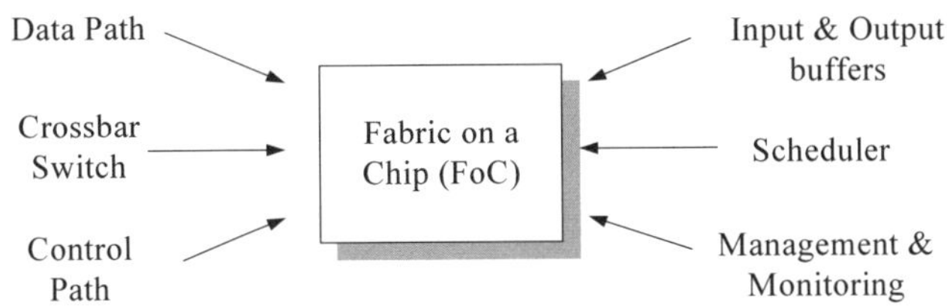

Figure 5.1. Potential functions to be consolidated as part of the FoC framework

products used in a switch or router. Lastly, the notion of FoC is coherent with the recent trend toward modern System on a Chip (SoC), a trend that is gaining momentum due to the inherent advantages it presents, in particular with respect to cost reduction.

Figure 5.1 illustrates the various components of a traditional input-queued switch; these components have the potential to be integrated on a single chip as part of the FoC framework. Improvements in the fabrication of VLSI circuitry play a key role as enablers for FoC. Due to advances in packaging technology, it becomes plausible to consider that all data packets arrive at the FoC directly. This reduces the need for virtual output queueing [3] and some output buffers associated with standard switch architectures. Due to the ability to embed multiple megabits of dual-port SRAM on a chip, packets can be efficiently stored and switched internally. We shall refer to packets as being of fixed size. This is generally true for all practical switch fabric designs, as external packets are typically segmented into fixed-size data units and reassembled as they exit the switch.

The crosspoint switches and scheduler, key components in input-queued switches, are avoided thereby substantially reducing chip count and power consumption. Correspondingly, much of the signaling and control information that typically spans multiple chips can be carried out on chip. Finally, the switch management and monitoring functions can be centralized since all the information is available at a single location.

5.2 Emulating an Output-queued Switch

Output-queued (OQ) switches offer several highly desirable performance characteristics, including minimal average packet delay, controllable Quality of Service (QoS) provisioning, and work-conservation under any admissible traffic conditions [4][5]. However, the memory bandwidth requirements of such systems is $O(NR)$, where N denotes the number of ports and R the data rate of each port. The latter is derived from the need to be able to accept (write) up to N arriving packets while simultaneously transmitting (reading) up to N departing packets. This requirement significantly limits the scalability of OQ switches with respect to their aggregate switching capacity. In an effort to mimic the desirable attributes of output-queued switches, while significantly reducing the memory bandwidth requirements, distributed shared memory architectures, such as the parallel shared memory (PSM) switch, have recently received attention [6]. In order to properly operate, the PSM switch must have sufficient bandwidth. At the core of distributed shared memory architectures is the memory management algorithm, which determines for each arriving packet, the memory unit in which it will be placed. However, the complexity of such algorithms found to date is O(N), where N denotes the number of switch ports, thereby inherently limiting the scalability of the scheme.

Initial work has indicated that, assuming each of the shared memory units can perform at most one packet-read or -write operation during each time slot, a sufficient number of memories needed to emulate a FCFS output queued switch is $K = 3N-1$ [6]. The latter can be proven by employing constraint sets analysis (also known as the "pigeon hole" principle), summarized as follows. An arriving packet must always be placed in a memory unit that is currently not being read from by any output port. Since there are N output ports, this first condition dictates at least N memory units are available. In addition, no arriving packet may be placed in a memory unit that contains a packet with the same departure time. This results in additional $N - 1$ memory units representing the $N - 1$ packets having the same departure time as the arriving packet, that may have already been placed in the memory units. Should this condition not be satisfied, two packets will be required to simultaneously depart from a memory unit that can only produce one packet in each time slot.

The third and last condition states that all N arriving packets must be placed in different memory units (since each memory can only perform one write operation). By aggregating these three conditions, it is shown that at least $3N - 1$ memory units must exist in order to guarantee FCFS output queueing emulation. Although this limit on the number of memories is sufficient, it has not been shown to be necessary. In fact, a tighter bound was recently found, suggesting that at least $2.25N$ memories are necessary [7]. Regardless of the precise minimal number of memories used, a key challenge relates to the practical realization of the memory management mechanism, *i.e.* the process that determines the memories in which arriving packet are placed. Observably, the above memory-management algorithm requires N iterations to complete.

In [8][9] Prakash, Sharif, and Aziz proposed the Switch–Memory–Switch (SMS) architecture as an abstraction of the M-series Internet core routers from Juniper. The approach consists of statistically matching input ports to memories, based on an iterative algorithm that statistically converges in $O(logN)$ time. However, in this scheme, each iteration comprises multiple operations of selecting a single element from a binary vector. Although the nodes operate concurrently from an implementation perspective, these algorithms are $O(log^2 N)$ at best (assuming $O(logN)$ operations are needed for each binary iteration as stated above). Since timing is a critical issue, the computational complexity should directly reflect the intricacy of the digital circuitry involved, as opposed to the high-level algorithmic perspective. It is important to note that in comparison to previous work, in particular that which addresses the SMS architecture, the FoC theme dictates that packet placement should be kept as simple as possible. Therefore, no crosspoint switches are to be used since they are very difficult to embed on chip, in particular at the high data rates considered.

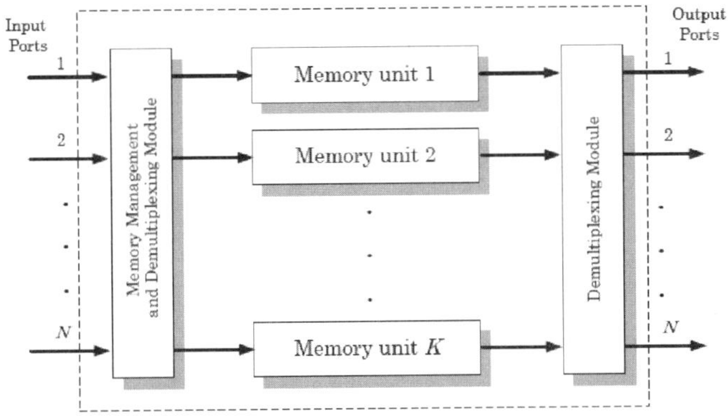

Figure 5.2. General structure governing the proposed parallel shared memory (PSM) switch architecture. Incoming packets are placed in a set of K ($>N$) memory units.

In light of the above material, the work presented in this chapter is motivated by the need for a packet-placement algorithm that is $O(logN)$ in speed and can utilize straightforward multiplexers to switch data segments from one location to another, as opposed to necessitating a crossbar switch. Acceptable costs, according to the paradigm fostered here, include fixed latency and a reasonable increase in the number of memory units used.

5.3 Packet Placement Algorithm

5.3.1 Switch Architecture

This section describes a proposed memory-management approach for PSM switches employing a pipeline architecture, as shown in Figure 5.2. The first step is to calculate the departure time of each arriving packet. Calculation of departure times is governed by the output scheduling algorithm used. The most straightforward scheduler is the first-come-first-serve scheme, in which packets are assigned departure times in accordance with their arrival order. To provide delay and rate guarantees, more sophisticated schedulers [4][5] can be referenced in order to determine the appropriate departure time assignment algorithm.

The main contribution here lies in the fact that the placement algorithm distributes the packet-placement process, thereby gaining execution speed at the cost of a fixed latency. In the proposed switch architecture, the memory-management algorithm is implemented using a multi-stage pipeline architecture, as depicted in Figure 5.3.

The pipeline architecture consists of $L \times L$ cell buffering units arranged in a square structure. Each row is associated with one of the parallel shared memory units. Hence, the architecture requires L parallel shared memories. Incoming packets

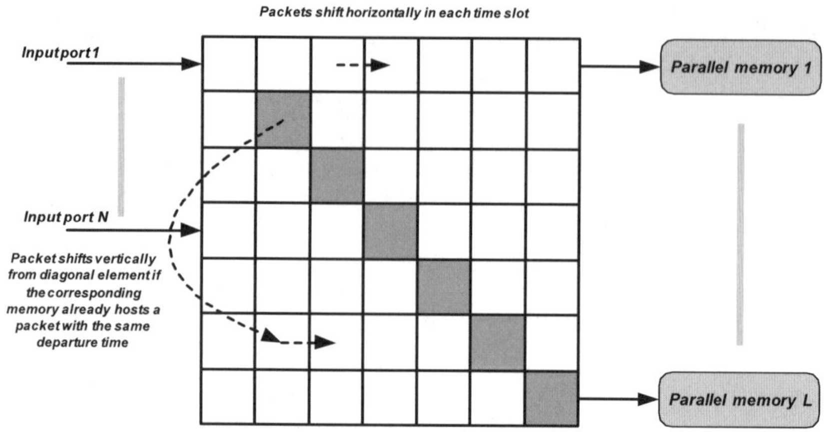

Figure 5.3. Memory-management pipeline architecture

from input port i are initially inserted into row i. The underlying mechanism is that at every time slot, packets are horizontally shifted one step to the right, with the exception of the diagonal cells of the structure. The diagonal cells have the ability to move vertically to another row of the same column. A cell moves vertically if any of the following two conditions are met: (a) the memory associated with the row in which it is currently located already contains a packet with the same departure time; (2) there is another cell ahead of it in the same row with the same departure time. Therefore, vertical moves are used as means of resolving memory placement contentions. The goal of the scheme is that once a packet reaches the last column of the pipeline, it is guaranteed to be located in a row which is associated with a memory that does not contain any packets with the same departure time.

5.3.2 Memory-management Algorithm and Related Resources

It should be apparent that each row, with the exception of the diagonal elements, can be considered as a simple shift register whereby packets are shifted one stage to the right at each time step. In each stage of the pipeline a single packet assignment is attempted per row. These assignment attempts are only made on the diagonal elements of the pipeline structure, so that there can be at most one assignment per column as well. The motivation for doing so is to isolate memory assignments, thereby reducing the complexity of the placement mechanism.

Each row maintains a mapping that specifies pre-allocated (or reserved) departure times of packets that have successfully passed the diagonal position. This mapping, which is essentially a binary mask, is referred to as the row's availability map. The diagonal element refers to this map in order to check if its departure time is available at that row. If it is available, the cell will horizontally shift to the right and mark the corresponding departure time in the availability map as reserved. If it is already reserved by another cell, it will shift vertically down to another row where the corresponding element position is empty and its departure time is not reserved.

Note that in this manner, following the first vertical jump, a cell may still have contention with (1) packets that are ahead of it with the same departure time but have not reached the diagonal element, or (2) packets having the same departure time that have leaped in the same time slot to the same row. To resolve contention originating from the first scenario mentioned, a second binary availability map called the reservation map is maintained. Upon shifting to a new row, each cell must update the row's reservation map by asserting the bit corresponding to its departure time. This allows future packets to guarantee that the row they are moving to does not contain packets that have arrived in a previous time step with the same departure time.

For practical reasons, both maps (the availability and reservation) cannot be unbounded in size. In practical switching systems, once a buffer reaches (or is close to reaching) its limit, flow-control signaling should be provided to the sources, indicating the need to either slow down or temporarily stop the flow of packets to a certain destination. Such a mechanism is always required since instantaneous data-traffic congestion may occur at any router or switch. In fact, even if the traffic is said to be statistically admissible, meaning that no input or output is oversubscribed, it may still be the case that for short periods of time a given output port, for example, is oversubscribed. To address such scenarios, and in an effort to reduce the probability of dropping any packets, the line cards typically host large memory spaces and traffic is regulated through the fabric.

In order to obtain an estimate of how much memory is needed, one can refer to the pure output-queued switch model. The approach taken is to obtain the expected queue size for each output port, from which a good approximation of the memory need is, in turn, obtained.

Consider an $N \times N$ output queue in which buffering occurs only at the output ports. During each time slot, the switch must transfer all arriving packets into their respective output ports. Let us first assume that arriving packet traffic is uniformly distributed across the output ports and obeys a (memoryless) Bernoulli i.i.d. process. Consequently, a packet arrives at a given input port with probability p and equal chance of being destined for any of the N output ports, *i.e.* uniformly and at random. Due to the inherent symmetry of the system, we shall analyze a single output port (*e.g.* output port 1) from which we will be able to derive the behavior at all other ports. Let Q_t denote the number of packets buffered in the output port buffer at time t. Since in each time slot, a maximum of N packets may arrive to a given output port and at most a single packet can depart, the following Markov model can be used to portray the evolution of the output buffer:

$$Q_{t+1} = Q_t + A_{t+1} - D_{t+1} \tag{5.1}$$

where $A_t \in [0, N]$ denotes the number of arrivals during time slot t and $D_t \in [0, 1]$ is an indicator function representing a departure event. Provided the queue size is not allowed to take negative values, then $D_t = 1$ if and only if $Q_t > 0$.. A_t is clearly i.i.d. (as defined above), and its distribution is given by

$$P(At = k) = \binom{N}{k} \left(\frac{p}{N}\right)^k \left(1 - \frac{p}{N}\right)^{N-k} \tag{5.2}$$

which converges to

$$\lim_{n \longrightarrow \infty} P\left(A_t = k\right) = e^{-p} \frac{p^k}{k!} \tag{5.3}$$

Moreover, since the arrival process is independent of the departure process, A_t is independent of Q_t.

Taking the expectation at both sides of (5.1), we consider the steady-state result by omitting the subscript t yielding the expectation expression $E[A] = P(Q > 0)$. However, we are primarily interested in $E[Q]$, so we square both sides of (5.1) and take the expectation, resulting in

$$E[Q^2] = E[Q^2] + E[A^2] + P(Q > 0) + 2E[AQ] - 2E[Q|_{Q>0}] - 2E[A|_{Q>0}] \tag{5.4}$$

Since A and Q are independent, we obtain

$$E[Q] = \frac{E[A^2] - E[A]}{2(1 - E[A])} \tag{5.5}$$

At this point, all that remains is to determine $E[A^2]$ and $E[A]$. In view of (5.3), we assert that $E[A] = p$ and $E[A^2] = p + p^2$, therefore

$$E[Q] = \frac{p^2}{2(1 - p)} \tag{5.6}$$

By multiplying the above term by N, we have the total average memory consumed by a pure output-queued switch. This provides us with a reasonable estimate of the amount of memory the PSM switch is to host on chip. As an example, for a load of 90% (*i.e.* $p = 0.9$) the per-port average memory requirements are merely 4.05 packets. The same analysis method may be applied to obtain the average queue size when bursty traffic is considered. It is widely acknowledged that real-life traffic tends to be correlated, or bursty, on many levels [10][11].

Consider a discrete-time, two-state Markov chain generating arrivals modeled by an ON/OFF source that alternates between the ON and OFF states. Let the parameters α and β denote the probabilities that the Markov chain remains in states ON and OFF, respectively. An arrival is generated for each time slot that the Markov chain spends in the ON state. Letting f_n denote the inter-arrival times distribution, the probability of two consecutive arrivals occurring is identical to the probability that following an arrival the Markov chain remains in state ON, *i.e.* $f_1 = \alpha$. Similarly, f_2 is the probability that following an arrival, the chain transitions to the OFF state and then returns to the ON state. For $n > 2$, it is apparent that following a transition from the ON state to the OFF state, there are $n - 2$ time slots during which the chain

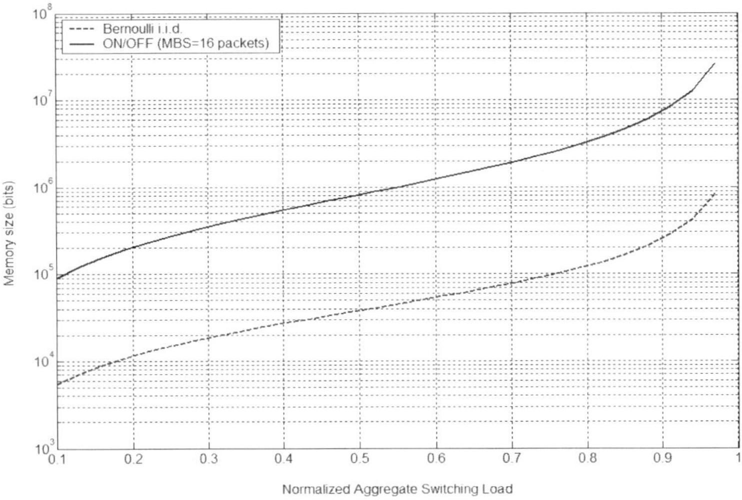

Figure 5.4. Average memory requirements of a FoC device

remains in the OFF state before returning to the ON state. Accordingly, we obtain the following general expression for f_n:

$$f_n = \begin{cases} \alpha & n = 1 \\ (1-\alpha)\,\beta^{n-2}\,(1-\beta) & n > 1 \end{cases} \tag{5.7}$$

From (5.7) the average burst size, $B = (1-\alpha)^{-1}$, and $E[A^2]$ can be directly obtained, from which we conclude that the average queue size is given by

$$E[Q_{(ON/OFF)}] = \frac{B\lambda(N-1)}{N(1-\lambda)} \tag{5.8}$$

where λ denotes the normalized average traffic load.

Note that in this case the average memory size is linearly proportional to the mean burst size. Figure 5.4 illustrates the aggregate amount of on-chip memory needed for a switch of 100 ports whereby packets are 64 bytes in size. The mean burst size is 16 packets. The reader should note that at 10 Gbps per port, this switch has an aggregate capacity of a terabit/s.

5.3.3 Sufficiency Condition on the Number of Memories

In this section, we provide some complexity results for the memory requirements of the pipeline architecture. In the proposed model, rows of the pipeline are arranged in P sequential blocks, whereby there is one row per input port (for a total of N rows) in the first block. As such, every input port writes its packets to one row. A cell in block r can only jump vertically to a cell in block $r+1$. In order to illustrate the underlying memory-management principal, we shall refer to the following example.

Example 1. Consider the simple scenario depicted in Figure 5.5. The state of the pipeline for four time slots (i.e. $t, t+1, t+4, t+5$) is shown. At time t, there are two packets in the second row and two packets in the third row, all with the same departure time D. Two packets are located at diagonal positions, and since there are two packets ahead of them with the same departure time, they shift down to a row in the second block (row 6) and then move one position to the right, as shown in the pipeline diagram for time $t+1$. Note that the packets were not permitted to move to row 5 since their position is already occupied with two other packets with departure times X and Y. At time $t+4$, the second packet with departure time D in row 6 reaches the diagonal element, and, since there is another packet ahead of it with the same departure time, it moves vertically to a row in the third block (row 7), followed by a one step shift to the right, as shown in the $t+5$ diagram. In this example, the pipeline consisted of three blocks of rows with 4, 2 and 1 rows in each, respectively. As will be discussed later, partitioning the structure into these blocks facilitates the complexity analysis as well as memory requirements of the architecture. We emphasize that the number of rows in each block is used only for illustration purposes. In the next lemma, we compute the number of the rows that are required for each block.

Lemma 1. *There should be at least* $2N-r$ *rows in block* r*, for* $r \in [2, 3, \ldots P]$.

Proof. Consider a cell that is shifting vertically from block r to block $r+1$. It will find at most $N-1$ rows blocked by other packets, since there have been at most $N-1$ other packets that may have arrived at the same time as this packet and may have shifted down in prior time slots. Moreover, it will find at most $N-r-1$ other rows having packets with the same departure time. Note that there could be at most $N-1$ other packets in the system with the same departure time; however, r of them have already been accounted for having caused this packet to jump to block $r+1$. Therefore, there are at most $2N-r-2$ rows (or locations) that this packet cannot move to in block $r+1$. Using the pigeon-hole principle, we conclude that $2N-r$ is a sufficient number of rows for block $r \in [2, 3, \ldots P]$.

From lemma 1, for a switch with N ports and P blocks, the total number of rows (parallel memories) can be expressed as

$$\begin{aligned} L(N) &= N + (2N-2) + (2N-3) + \ldots + (2N-P) \\ &= N + 2N(P-1) - (P+2)(P-1)/2 \\ &= 2(P-1)N - (P+2)(P-1)/2 \end{aligned} \tag{5.9}$$

In order to compute the number of rows, we have to find $P(N)$, which directly reflects the maximum number of vertical shifts a packet can perform prior to being successfully assigned to a memory. We shall refer to Example 1 and derive an upper limit on the number of conflicting packets with the same departure time in the fourth

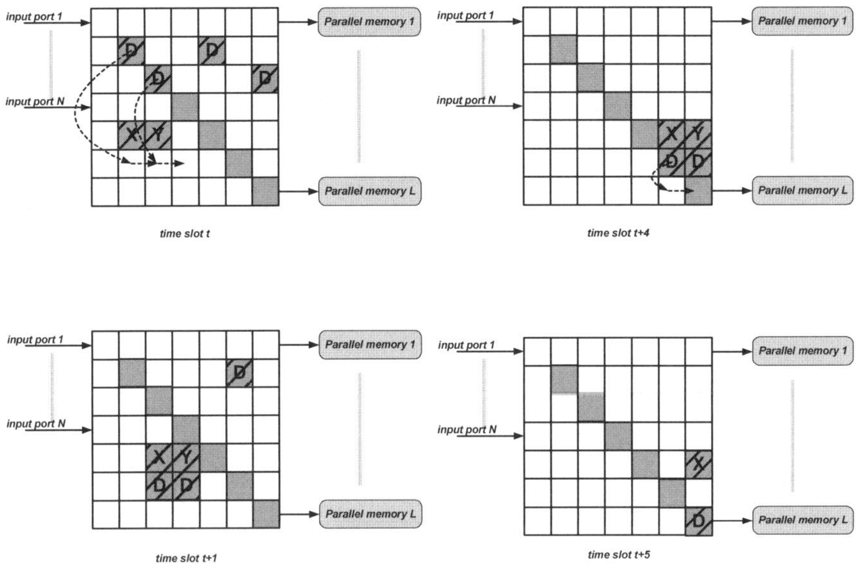

Figure 5.5. Example illustrating the proposed memory-management algorithm for a 4-port switch. The state of the pipeline structure is depicted for 4 consecutive time slots

block, *i.e.* after three vertical shifts have occurred. We would like to find an upper limit on the number of conflicting packets after three shifts and use that recursively to obtain the maximum number of jumps, and hence blocks, that are sufficient to guarantee that each arriving packet will be successfully assigned a memory.

Lemma 2. *The maximum number of packets with the same departure time in the fourth block is* $(\sqrt(N) - 1)^2$.

Proof. Suppose that there are N_1 packets with the same departure time in the first block. Throughout the proof, we shall refer to this set of packets having the same departure time. Note that $N_1 \leq N$, since there cannot be more than N packets with the same departure time in the system. Let us assume that these packets are located in M_1 ($M_1 \leq N$) rows of the first block. Therefore, the number of packets that move vertically to the second block will be $N_2 = N_1 - M_1$. Next, we compute the number of rows in the second block that contain one of these packets. There are $N_2 = N_1 - M_1$. packets that have moved to the second block, and the maximum number of packets that can shift simultaneously to the same row is M_1. Hence,

$$M_2 \geq \left\lfloor \frac{N_1 - M_1}{M_1} \right\rfloor \tag{5.10}$$

Therefore, the maximum number of packets with the same departure time that can move to the third block is given by

$$N_3 = N_2 - M_2 \leq N_1 - M_1 - \left\lfloor \frac{N_1 - M_1}{M_1} \right\rfloor \tag{5.11}$$

If $N_1 - M_1$ is divisible by M_1, then

$$N_3 \leq N_1 \left(1 - \frac{1}{M_1}\right) - M_1 + 1 \tag{5.12}$$

otherwise, since $N_4 \leq N_3 - 1$, we have

$$\begin{aligned} N_3 &\leq N_1 \left(1 - \frac{1}{M_1}\right) - M_1 + 2 \\ N_4 &\leq N_1 \left(1 - \frac{1}{M_1}\right) - M_1 + 1 \end{aligned} \tag{5.13}$$

The maximum value of the expression in (5.13) is reached when $M_1 = \sqrt{N_1}$. Substituting $N_1 = N$, yields the inequality

$$N_4 \leq \left(\sqrt{N} - 1\right)^2 \tag{5.14}$$

Note that if N is a complete square we have,

$$N_3 \leq \left(\sqrt{N} - 1\right)^2 \tag{5.15}$$

In the following corollary, we exploit these results to determine the order of the memory blocks.

Corollary 1. *A sufficient number of parallel memory blocks required for an $N \times N$ switch, employing the proposed architecture, is $O\left(\sqrt{N}\right)$.*

Proof. Equation (5.14) shows that for an N-port switch, the maximum number of conflicting packets with the same departure time in the fourth block is at most $N - 2\sqrt{N} + 1$. Let $P(N)$ represent the number of stages required for an N-port switch. We thus have

$$\begin{aligned} P(N) &= P(N - 2\sqrt{N} + 1) + 3 \\ P(1) &= 1 \end{aligned} \tag{5.16}$$

from which we conclude that $P(N) = O\left(\sqrt{N}\right)$.

Theorem 1. *For an N-port switch, where $N \leq k^2 k \in \{1, 2, \ldots.\}$, the number of memories is*

$$L(N) \leq 4k^3 - 5k^2 + k + 1 \tag{5.17}$$

with equality if $N = k^2$.

Table 5.1. Breakdown of the number of rows in each block for a 16-port PSM switch

Block Number	Number of Memories
1	16
2	30
3	29
4	28
5	27
6	26
7	25

Proof. We prove the equality for $N = k^2$, suggesting that the general case trivially follows. We first show by induction that the number of required row blocks are $P(k^2) = 2k - 1$. For $k = 1$ the result is trivial. We assume that the result holds for k and infer it for $k + 1$. For $N = (k + 1)^2$, using Lemma 2 and (5.15) (given that N is a complete square), we have

$$P\left((k+1)^2\right) = P\left(k^2\right) + 2 = 2k - 1 + 2 = 2(k+1) - 1 \tag{5.18}$$

Then, we utilize the relationship proven for L in (5.9) to obtain

$$\begin{aligned} L\left(k^2\right) &= (2P-1)N - (P+2)(P-1)/2 \\ &= (4k-3)k^2 - (2k+1)(2k-2)/2 \\ &= 4k^3 - 3k^2 - 2k^2 + k + 1 \\ &= 4k^3 - 5k^2 + k + 1 \end{aligned} \tag{5.19}$$

The above theorem states that the number of parallel memories required in this architecture is $O(N^{1.5})$. Given that the minimum number of memories is $O(N)$, the increased memory requirement observed is the price paid in order make the assignment problem parallel and feasible from an implementation perspective. For example, for$N = 16$ (*i.e.* $k = 4$), the number of parallel memory elements is 177, arranged in 7 blocks. Table 4.1 provides the number of rows in each block for such a switch.

Lemma 2 provides an upper bound on the number of memories required. In order to illustrate a case whereby this is reached, we shall refer to the following example.

Example 2. We refer to the following adversarial scenario pertaining to a 9-port switch, which is easily extendable to larger switch sizes when N is a complete square. Consider the settings illustrated in Figure 5.6. There are nine packets with the same departure time residing in the first block of memory. The first three packets are scheduled, while the others are shifted vertically to the second block. Note that packets 4, 5 and 6 reach the diagonal together and hence move down simultaneously. As shown, they may end up in the same row. A similar pattern of behavior is observed for packets 7, 8 and 9. Packets 6 and 9 are scheduled in the second block, while packets 5, 8 and 4, and 7 move to the third block. Once again packets 5 and 8 move simultaneously to the same row and packets 4 and 7 to a different row. Clearly, after two moves we end up with a set of four conflicting packets, *e.g.* from $k = 3$ we reach a condition with $k = 2$. Similarly, we need two more shifts to reach a point with only one packet in a block.

5.4 Implementation Considerations

5.4.1 Logic Dataflow

This section provides a detailed discussion on the different implementation aspects pertaining to the proposed architecture. In particular, we focus on the timing requirements, using a practical realization of the proposed scheme, and derive related scalability properties.

It should be evident that the critical path in the design is the memory assignment process that takes place at the diagonal elements, or decision cells, of the pipeline structure. Let us review the assignment process. Each decision cell, residing on row r_i $(i \in [1, 2, \ldots L])$ must determine whether a packet, p_i, can be placed at the memory associated with row i. The memory is considered available for packet placement if it does not already contain a packet with the same departure time d_i as the packet residing in the decision cell. If the memory is occupied with a packet with the same departure time, then the packet is shifted to another row.

The following three criteria govern the movement of packets to a new row r_n:

1. The memory located on the selected row (r_n) cannot contain a packet departing at time d_i.
2. No other packets on row r_n, ahead of the packet moved and which have yet to reach a diagonal location, can have a departure time d_i.
3. Row r_n must not contain one of the $N - 1$ possible packets that have the same arrival time as the shifted packet. In other words, the position in the new row should be unoccupied.

To satisfy the first placement rule, each memory maintains an availability vector (see Section 5.3.2) of length k, to indicate locations available for the placement of a packet at time d_i. In a distributed shared memory switch, each shared memory will contain k cells representing k consecutive departure times. This dictates the size of the availability map (see Section 5.3.2) which consists of L availability vectors. Note that in order to perform the appropriate selection for packet p_i with departure time d_i, the column of bits pertaining to the dth position in the availability map should be examined. Based on Lemma 1, since a packet always shifts into the next block, we note that the maximal number of rows that is to be examined by each diagonal location for shifting a packet is $2N - 2$. This inherently bounds the critical path of the design, since it defines an upper bound on the search space that must be considered at each diagonal element.

While the availability map provides information as to the memories contents, it does not provide any information regarding the location of packets that arrived

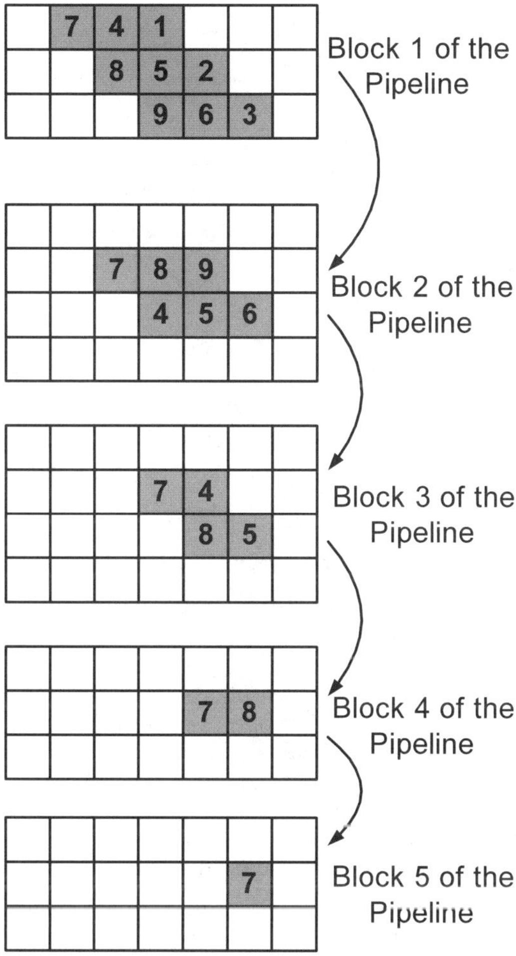

Figure 5.6. Adversarial scenario demonstrating the sufficiency bound on the number of memory units for a 9-port switch

during the same time slot as p_i. There are at most $N-1$ packets that arrived at the same time as p_i. Having stated that a vector of length $2N-2$ is sufficient to locate an available row, an occupancy vector must be created to determine the location of $N-1$ packets within the $2N-2$ subset reflecting potential adequate rows. Hence, the bitwise-OR of the occupancy vector and the availability vector provides a single decision vector representing viable rows for packet placement. The decision vector defines rows which (1) are not associated with memories that contain a packet with departure time d_i, and which (2) do not contain a packet with the same arrival time as p_i.

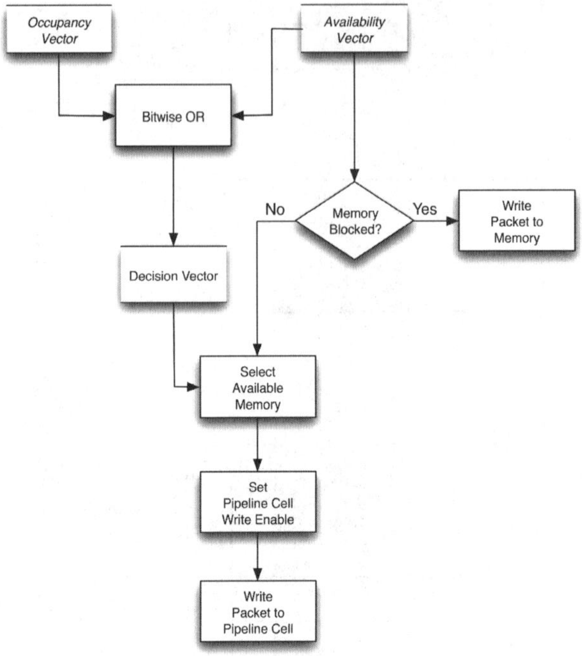

Figure 5.7. Outline of the process performed at the decision cells

The memory-management algorithm is based on selecting an available memory from one of $2N - 2$ memories, reflected by the decision vector, such that all N packets are placed at the end of the $O(N^{1.5})$ phases. The decisions made at each location are represented by the flowchart in Figure 5.7.

As stated previously, we first determine whether the packet in the decision cell is blocked by a packet with the same departure time as p_i. If the least significant bit (LSB) of the availability vector, representing the memory unit residing on the decision cell's row, is set to '1', then the memory is considered blocked and the packet must be placed on a different row. In this case, the decision cell uses the result of the bitwise OR between the availability and occupancy vectors to create the decision vector. This decision vector is then presented to a row selection unit, which selects an available row from one of $2N - 2$ rows. Once the row is selected, the decision cell presents a write enable, as well as p_i, to each of the $2N - 2$ pipeline packets below the current decision row during the next pipeline stage. The result of this decision process is that p_i is either written to the memory associated with the row in which it resides or placed in a row that is free of conflict.

To this point, the availability vector has been presented as a bitmap of memories with a vacancy at departure time d_i. It should be noted that although each decision cell only requires an availability vector of size $2N - 2$, the availability map consists

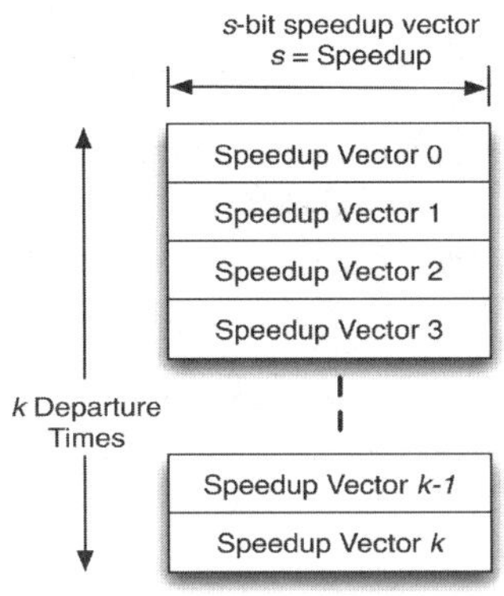

Figure 5.8. Illustration of the memory speedup logic representation

of a set of L vectors since each decision cell is offset by one memory relative to the decision cell located on the prior row. Moreover, it is unknown at which departure time a decision cell will request, further establishing the need to maintain a $k \times L$ availability map. A k-to-1 multiplexor is used to select one of the L bit vectors with a $2N - 2$ bit-sliced availability vector to be presented to each decision cell.

In order to avoid an N^2 memory requirement, we must refrain from placing packet p_i in a row that already contains a separate packet with the same departure time as p_i. To support this requirement, packets must update the availability map when they are placed in a new row. It has been stated that decision cells place conflicting packets in one of $2N-2$ rows below it. Given that $2N-2$ decision cells could select the same row, a logical OR of the $2N - 2$ potential departure times located in each memory's row is required to maintain the integrity of the availability map. If we maintain a strict row-to-physical-memory coupling, we can introduce speedup into the system. Let the memory speedup s be defined as the number of logical memories assigned to a single physical memory, for each departure time.

We can now also state that a packet can be placed in a physical memory, at a given departure time, so long as there exists at least one logical memory location available at that departure time for the physical memory corresponding to the packet's row. This allows us to view the availability of a single memory as an $s \times k$ register file, illustrated in Figure 5.8.

Furthermore, we can assert that a decision cell only requires information that indicates at least one logical memory location is available for a given departure time. Considering such simplification, we can directly apply a bitwise OR to each speedup vector at a given departure time, as presented in Figure 5.9, to derive a single coherent availability bit.

This allows us to represent available memories to the decision cells from a physical perspective, rather than a logical one, resulting in a decreased decision time. The delay cost of a logical-to-physical memory address translation is only introduced

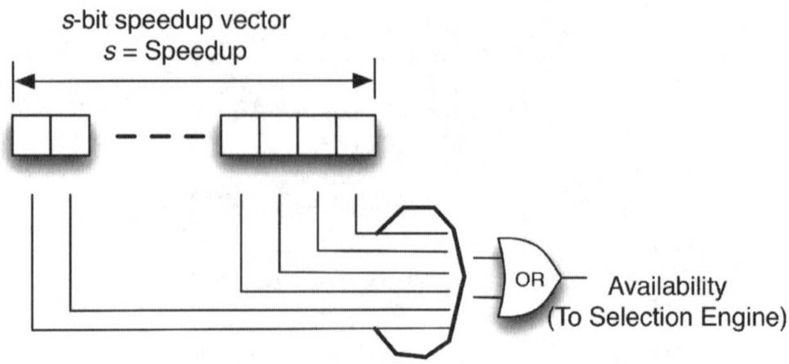

Figure 5.9. Illustration of the memory speedup logic representation

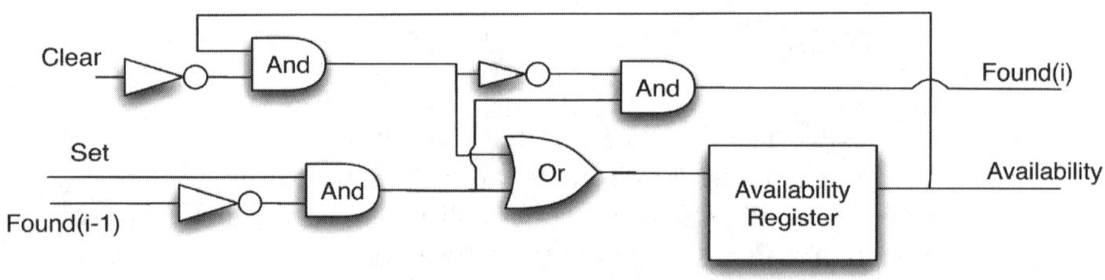

Figure 5.10. Consolidation of the speedup vector to derive the available row location

when the decision cell actually places the packet. In effect, the burden of placing packet p_i at one of s offsets at departure time d_i is shifted from the decision cell to the memory unit. The departure time serves as the base address for writing a packet to that memory. When speedup exists, the offset address within a given departure time is determined when a memory occupation bit is set in the single memory availability vector. The single memory availability vector is constructed from availability cells, the logical building blocks, presented in Figure 5.10, implement a ripple structure to set a single register within a speedup vector. The decision cell drives the set signal high for a given departure time. The set signal is then distributed to all availability cells within a given speedup vector. A found signal is propagated from the least significant bit to the most significant bit of the speedup vector to indicate that an available offset has been located.

The first availability cell to locate an available offset blocks all subsequent availability cells from setting their availability cell bit. The inherent delay in ripple structures is not considered to be significant, given that speedup vectors are relatively small with lengths ranging from 2, 4, or 8 bits wide. The alternative to the ripple delay is allowing the decision cells to select an available memory from $(2N-2)*s$ logical memory locations. Given this alternative, the ripple delay is preferable.

While we have covered all aspects of the placement process, we have yet to address issues associated with retrieving packets from the shared memory structure and transmitting them to the correct output, or egress, port. Given a set of $O(N^{1.5})$ memories, constructing a 1-to-$O(N^{1.5})$ multiplexor at each output port is impractical for the desired port densities. A more practical solution is to present a single shared packet bus to each shared memory unit. As packets are read from memory, each egress port will have at most one packet destined for it at any given time. Correspondingly, we can present the shared packet bus, of size $N * packet\_size$, to each shared memory unit such that each jth bus slice, of size *packet_size*, represents a packet destined for output j. Each shared memory unit determines the destination j of the packet it is transmitting to an egress port and only drives the jth bus slice of the packet bus. As such, each egress port will only be responsible for transmitting the packet located at the jth bus slice, corresponding to the port enumeration of the packet bus. The implementation cost of this approach is reflected in the requirement for an N-to-1 multiplexor to be located in each memory cell.

5.4.2 FPGA Implementation Results

To establish the viability of the proposed multistage PSM memory-management architecture, we have implemented the critical path in VHDL and synthesized its components targeting a Xilinx Virtex-4 XC4VLX200-10-FF1513 device. This implementation consists of 16 ports operating at 40 Gbps each, representing a switch with an aggregate capacity of 640 Gbps. The maximum departure time value configured to be 64. With no logic speedup, the physical resource requirements for the system are as follows: physical memories – 176, decision cells – 176, registers $\sim= 15.4k$, pipeline depth – 176. After packets are placed into shared memories, they are presented at the output of the PSM switch every 12.5 ns. Given a link speed of 40 Gbps, and considering 64-byte packets, the scheme is only realizable if packet placement decisions can be made in less than 12.5 ns. We therefore consider the worst-case path, along with associated delays attributed to each decision made along this path.

First, we have identified the worst-case path origin to be the availability vector, which is obtained as an output of a k-to-1 multiplexer applied to the availability map. The maximum departure time value, having previously stated to be 64, requires the k-to-1 multiplexer design multiplex 64 potential availability vectors for presentation at the decision cell. The resulting 64-to-1 multiplexer was found to have an overall delay of 2.57 ns.

The selected availability vector is then used by the decision cell to determine the correct row in which to place conflicting packets. This decision process directly corresponds to the flow diagram presented in Figure 5.7. The time required to identify a memory conflict, followed by locating an available row, was found to be 3.823 ns. Once the available row is located, a bitwise OR is applied to the departure time and the potential $2N - 2$ packets that may be attempting to relocate to the same row. This is required, as discussed earlier, to maintain the integrity of the availability map. The delay associated with this wide OR was 1.52 ns. Registering of the departure time denotes the terminating point of the worst-case path, from which we can assert that the worst-case delay is 7.913 ns – well below the 12.5 ns limit stated above. As a final note, the overall latency contributed by the architecture with respect to a pure output-queued switch is 2.2 μs (176 stages of 12.5 ns each).

5.5 Conclusions

The notion of designing a packet switching fabric on a chip was introduced and studied from a pragmatic as well as technological perspective. It has been argued that in the context of emulating an output-queued switch, a core challenge pertains to the memory-management algorithm employed. A proposed packet-placement algorithm and related architecture were described in detail, emphasizing the feasibility attributes. The switch model and framework presented can be broadened to further investigate the notion of consolidating multiple switch fabric functions on silicon.

References

1. I. Elhanany, D. Chiou, V. Tabatabaee, R. Noro, and A. Poursepanj. The network processing forum switch fabric benchmark specifications: An overview. *IEEE Network Magazine*, 19(2):4–9, 2005.
2. Cisco systems crs1 multishelf router. available at: www.cisco.com.
3. [Y. Tamir and G. Frazier. Higher performance multiqueue buffers for vlsi communication switches. In *15th Annual Symposium on Computer Architecture*, pages 343–354, 1988.
4. A. Parekh and R. Gallager. A generalized processor sharing approach to flow control in integrated services networks: The single node case. *IEEE/ACM Transactions on Networking*, 1(3):344–357, June 1993.
5. M. Shreedhar and G. Varghese. Efficient fair queueing using deficit round robin. *Proc. of ACM SIGCOMM '95*, pages 231–242, September 1995.
6. R. Zhang S. Iyer and N. Mckeown. Routers with a single of buffering. *ACM Computer Communication Review (SIGCOMM'02)*, pages 251–264, 2002.
7. H. Liu and D. Mosk-Aoyama. Memory management algorithms for dsm switches. *Stanford Technical Paper*, July 2004.
8. A. Prakash A. Aziz and V. Ramachandra. A near optimal scheduler for switch-memory-switch routers. In *Proceedings of the Fifteenth Annual ACM Symposium on Parallel Algorithms and Architectures*, pages 343–352, July 2003.
9. A. Prakash, A. Aziz, and V. Ramachandra. Randomized parallel schedulers for switch-memory-switch routers: Analysis and numerical studies. IEEE INFOCOM 2004, 2004.
10. C. Williamson. Internet traffic measurement. *IEEE Internet Computing*, (5):70–74, 2001.
11. C. Barakat, P. Thiran, G. Iannaccone, and C. Diot. On internet backbone traffic modeling. ACM Sigmetrics, 2002.

6

Packet Switch with Internally Buffered Crossbars

Zhen Guo, Roberto Rojas-Cessa, and Nirwan Ansari

New Jersey Institute of Technology (NJIT), Department of Electrical and Computer Engineering. {zhen.guo,rojasces,nirwan.ansari}@njit.edu

The increasing demand for higher data rates on the Internet requires routers that deliver high performance for high-speed connections. While output-buffered switches are known for their limited scalability, input-buffered (IB) switches have been of interest for research and commercialization. However, IB switches may not be able to keep up with the increasing data rates as optical technology advances rapidly. To keep up with the increasing data rates, switches based on internally buffered crossbars are considered a feasible solution for the next generation packet switches. The cost associated with internally buffered crossbar switches is tied to the addition of buffers to the crosspoint elements. Here, we present an overview of several internally buffered crossbar switches and their selection schemes. Furthermore, we discuss the properties of the different approaches to internally buffered packet switches.

6.1 Introduction to Packet Switches

The exchange of information within the Internet is made possible by the switches and routers interconnecting networks. These routers facilitate networks to communicate by using a shared language or protocol. An example of such protocols employed at different network layers are Ethernet, Asynchronous Transfer Mode (ATM), or the popular TCP/IP suite. These protocols determine the packet formats and the way to find a route from the source host to the destination host.

In the remainder of this chapter, we consider the IP protocol to define the packet term. Therefore, switches and routers are required to process IP packets. However, a large number of packet switches base their architecture on ATM technology, where, different from IP packets, ATM packets have fixed lengths, called cells. IP packets, however, can be handled by cell-based switches as variable-length packets are segmented at the input ports, and switched from input to output in a cell-based fashion. Variable-length IP packets are re-assembled at the output ports before they depart to other switches. Here, we refer to cells as fixed-length packets, which are not necessarily ATM cells.

Packet switches identify the destination of packets at the input ports and forward them to the appropriate output ports, completing in this way the packet processing at layer 2 of the Open System Interconnection (OSI) model. Those switches that find out information about the connectivity between networks and paths to reach different possible destinations are referred to as routers. This information is summarized in a forwarding table that is used to determine the output port of the switch according to the destination of the traversing packet. These routers are devices performing tasks at layers 2 and 3. Once output ports are defined in a forwarding table, a switch performs the scheduling and forwarding of packets.

As interconnection technologies mature, such as those based on optical technology, data rate increases and routers need to keep up with that by processing packets fast. The functions that require high performance in routers are identifying the packet type, so that the packet can be processed accordingly (including forwarding), and switching the packet from an input to an output port. Here, we focus on switching a packet from the input port to an output port. The task seems simple; however, it gets complex as there is the possibility that several packets need to go from different inputs to the same output. Therefore, this creates the necessity of interconnecting input and output ports, buffering packets and scheduling the packet switching time. A scheduler selects the time a packet is switched to the output. The complexity of the scheduler depends on the buffering strategy and on the selection scheme used.

A packet switch is comprised of input port cards and a switch fabric. An input port card, also known as a line interface card (LIC), determines the processing the switch performs on each packet. Some of these functions are: packet classification, destination lookup (IP lookup for IP packets), buffering, packet modification, and packet interfacing for internal switching. The switch fabric is an interconnection network used by the LICs. The way a switch operates internally depends on the switch fabric used. A plethora of switch fabrics have been developed for packet switches. Many of them were developed for the telephone network and are now applied to packet networks. We can divide switches into single- and multiple-stage. A single-stage switch is a fully interconnected network, where all inputs are directly connected to each output. Crossbar switches are examples of single-stage switches. In a multiple-stage switch (such as Clos [1] and Banyan [2] networks), each input is connected to a switch module instead of to all other outputs, and a packet may reach its destination after passing through two or more switch modules. A switch module is a small switch with two or more ports. A good example of a multiple stage switch is the 3-stage Benes switch.

6.2 Crossbar-based Switches

Crossbar switching fabrics are very popular for switch implementation due to their non-blocking capability, simplicity, modularity, and their market availability. In a crossbar, there are N^2 switch elements, also called crosspoints, for N input ports and N output ports allowing any input to connect to any output. The performance of a switch depends on the adopted buffering strategy.

An OQ switch has queues at each output. All arriving cells at the inputs must be immediately delivered to their outputs, and therefore, the output queues must be able to store up to N cells in a time slot. This requires high interconnection and memory bandwidth at a speedup of N times the line rate. For an $N \times N$ OQ switch, the memory must be able to accommodate N write accesses (to write N cells into output buffers) and one read access (to send one cell to the outgoing link) in one-cell time. This requirement is known as internal speedup of a switch (defined as the number of times that the switch core works faster than the input line rate). Unfortunately, building memories with suitable working speed for a moderate speedup is prohibitively expensive. However, the OB switch exhibits the optimum behavior of an ideal switch and therefore, it has been used as a benchmark for comparison of switch performance.

A switch with buffers[1] at the inputs is termed an input-buffered (IB) switch. IB switches are desirable because of their scalability and low hardware requirement. IB switches have an internal speedup of 1 (also considered as no speedup) because the crossbar fabric has the same speed as that of the external lines. It is well known that, if first-in first-out (FIFO) input queues are used to hold arriving packets, at every time slot only the head-of-line (HOL) cell is considered. When a HOL cell cannot be cleared due to its loss in output contention, it may block every cell behind that may be destined for a currently idle output. This phenomenon is called the HOL blocking problem, which limits the throughput to only 58.6% [3]. To eliminate HOL blocking, virtual output queuing (VOQ) can be used; each input buffer is partitioned into N queues with one queue for each output port. Each arriving packet is classified and then queued into the appropriate VOQ according to its destination output port.

However, IB switches [4] need to resolve input and output contentions before cells are forwarded to the outputs. Arbiters at input and outputs perform contention resolution by means of a matching process. Furthermore, the switching performance of an IB switch requires complex matching schemes to provide high-performance switching. This high complexity limits switch port speeds. The requirements for arbiters to be feasible and to provide a high performance are: (a) low complexity, (b) fast contention resolution, (c) fairness, and (d) high matching efficiency. As an example, a matching scheme must perform input or output arbitration within 8 ns in an IB switch with 40 Gbps (OC-768) ports and 80-byte cells, assuming that input and output arbitrations may use up to half a time slot and that the transmission delays are decreased to negligible amounts (*e.g.* the arbiters are implemented in the same chip, in a centralized way).

A matching can be maximum or maximal. A maximum match is a maximum cardinality bipartite matching of input with packets queued to N outputs. A maximal match is a matching that cannot be improved without removing some input–output matches. Theoretically, a maximum weight matching (MWM) algorithm provides 100% throughput for any no-overbooking traffic. However, the scheme's complexity prevents its implementation for fast speeds. Maximal matching schemes have been considered as an alternative to maximum matching schemes; iSLIP, dual round-robin

[1] This chapter uses the terms queue and buffer interchangeably.

matching (DRRM), and longest output occupancy first algorithm (LOOFA) are some examples. To make up for the lack of efficiency that a maximal scheme has (compared to a maximum type), a number of iterations, where the number of iterations is the number of times that an algorithm is performed to obtain a cumulative result, speedup, or the number of both is used, as in LOOFA. iSLIP is a typical example of an iterative matching scheme. iSLIP provides 100% throughput for uniform traffic, but because of the arbitration time, it has been proposed for a small number of ports due to its centralized implementation. Transmission of phases such as request, grant and acknowledge are performed within a cell slot between input and output arbiters. This transmission of information reduces the available time for arbitration because transmission phases are performed during the cell slot in serial with input and output arbitration, even when the transmission is done within a single chip.

Although Store-Sort-and-Forward (SSF) [5], a frame-based scheduling scheme for IB switches, has proven to provide QoS guarantees without requiring speedup, speedup is another approach to tackle the lack of efficiency by scheduling in IB switches. When the internal speedup is between 1 and $N - 1$, buffering is required at both the inputs and outputs. Hence, a combination of an input-buffered and an output-buffered switch is required, which is a combined input output buffered (CIOB) switch. With a speedup of 2, most maximal matching algorithms can achieve 100% throughput under any admissible traffic. However, as the demand for high switching rates increases, the speedup shortens the time for performing matching because this time is divided by speedup. Therefore, iterative-based algorithms may have no time to find a maximal matching and neither to perform several iterations.

6.3 Internally Buffered Crossbars

Internally buffered crossbar switches employ buffers within the switch elements. In these switches, packets go from an input line to the crosspoint buffers and wait there until they are served towards the output port. Therefore, input selection is performed at each output of the buffered crossbar. We call a buffered crossbar (BC) switch a pure buffered crossbar because in this architecture buffering is only at the crosspoints. When more than one cell is contending for the limited link capacity at its outputs, buffers store the cells that lose contention. Since the buffer size is finite, arriving cells are discarded when the buffer becomes fully occupied.

Since the number of buffers in a crossbar grows in the same order as the number of crosspoints, $O(N^2)$, the implementation was considered costly for a large buffer size or large N in which VLSI memory implementation requires a large amount of real estate in a chip. As the technology for chip manufacture has matured, the implementation of buffered crossbar switches has become feasible.

One of the first pure buffered crossbar switches was proposed in [6], and later in [7]. However, this architecture was named the Butterfly switch at the time.

Another of the first internally buffered switches was presented in [8], where a 2×2 crossbar chip with a crosspoint buffer size of 16 Kbytes, with input buffers.

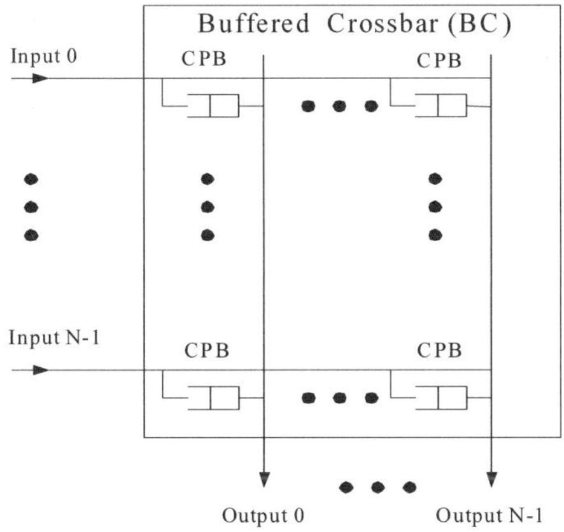

Figure 6.1. Crosspoint buffered switch

The output polls the crosspoint buffers to decide which packet can be served out of the buffered crossbar.

Buffered crossbar switches need output arbiters to select a cell, stored in the crosspoint buffer, to be sent to the output port. Therefore, the performance of this switch depends on the output arbitration. The output arbitration scheme considers all crosspoint buffers and selects one according to a selection policy. Therefore, the time and computation complexity of an output arbitration scheme is a function of the number of crosspoint buffers (or inputs, N) at an output, or $O(N)$.

An example of a BC switch that emphasizes output arbitration is the scalable distributed arbitration (SDA) switch [9]. The SDA switch reduces arbitration time by using a distributed approach for the output arbitration in each output. Instead of considering N inputs at any given time, an arbiter is partitioned into $N-1$ selectors, where each selector considers two buffers. This switch performs random selection of cells (crosspoints) at each output. The SDA switch has a crosspoint buffer, a transit buffer, an arbitration-control block (CNTL), and a selector at every crosspoint. A crosspoint buffer sends a request (REQ) to CNTL if there is at least one cell stored in the crosspoint buffer. A transit buffer stores several cells that are sent from either the upper crosspoint buffer or upper transit buffer. The transit buffer has a size of one or a few cells. The transit buffer size is determined by the round-trip delay of control signals between two adjacent crosspoints. The longest control signal transmission distance for arbitration within one cell time is the distance between two adjacent crosspoints. In a switch with implementation of the output arbiters in a centralized fashion, the control signal for arbitration must pass through all the crosspoint buffers belonging to the same output line to complete the selection each time slot, and therefore, the arbitration time depends on the number of inputs (or crosspoints). In this way, the arbitration time in the SDA switch is independent of the number of input ports.

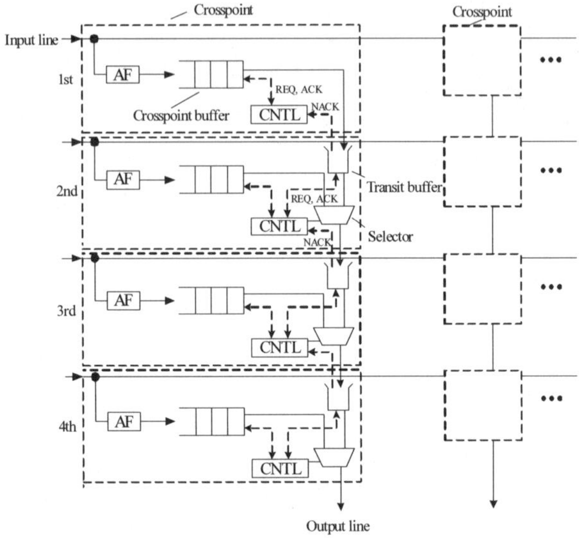

Figure 6.2. Scalable distributed-arbitration switch structure

The SDA switch was tested under uniform traffic with Bernoulli arrivals with an input load of 0.95 for different switch sizes ($N = \{4, ..., 32\}$). This switch delivers a cell average delay of less than 100 time slots. Another feature of the SDA switch is fairness. This is achieved by the nature of the distribute selection scheme using different selection probabilities for different inputs at a crosspoint, such that the total selection probability by an output is the same for any input.

6.4 Combined Input–Crosspoint Buffered (CICB) Crossbars

In order to reduce the crosspoint buffer size, input buffers can be used with larger capacity as these buffers are located in the input ports, and the amount of memory at the input ports (outside of the buffered crossbar) can be of large size (*e.g.* several memory chips). CICB switches that use the FIFO policy in their input buffers are the simplest of them, and are called FIFO–CICB.

6.4.1 FIFO–CICB Switches

In [10], an input and crosspoint buffering matrix switching architecture with FIFO input buffers, or FIFO–CICB switch, was proposed. In [11], a FIFO–CICB switch input buffers and random selection policy at the output was proposed and shown to provide high throughput. CICB switches with single-cell crosspoint buffers were proposed in [12, 13]. These switches also used FIFO input buffers at the input ports,

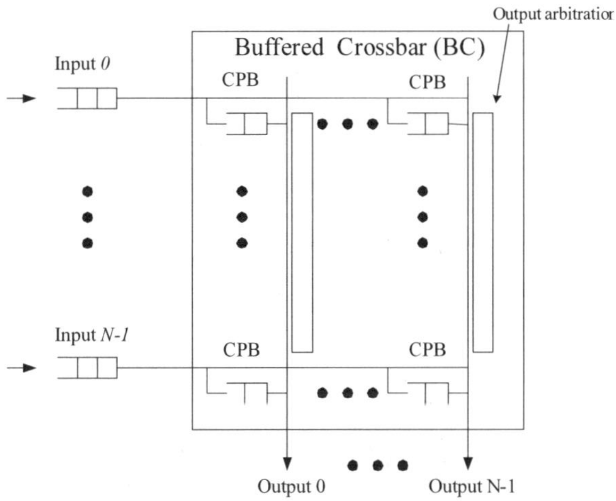

Figure 6.3. Combined input–crosspoint buffered crossbar switch with FIFO input buffers

or FIFO–CICB switches. The switches provide a throughput of 91%, where, however, the HOL blocking [3] was still present. The FIFO buffers at the inputs limit the maximum throughput in that switch because the HOL blocking cannot be completely eliminated. These switches showed the need to remove the HOL blocking that was present in IB switch and then, inherited by FIFO–CICB switches.

Another example of a FIFO–CICB switch, with a different architecture approach, a multiple-plane architecture, is the Tandem-crosspoint (TDXP) switch [14]. The main purpose of the TDXP switch is to overcome HOL blocking by using the parallel switch technology. This switch has multiple crossbar switch planes as shown in Figure 6.4. The switch planes are connected in tandem at each crosspoint. There is a one-cell buffer in the crosspoint. The internal speedup in each plane is the same as the input/output line speed. Each switch plane can transmit only one cell to each output port within one cell time slot. The HOL blocking phenomenon occurs at the input buffers, where the FIFO policy is used.

The TDXP switch improves the switching performance by using multiple planes. In this way, cells will not cause HOL blocking as there is room for a cell in each switch plane. This is similar to letting the first K cells of a FIFO participate in the output arbitration process, where K is the number of planes in the TDXP switch. In a 32×32 switch, the throughput provided is above 95%, which is a significant improvement over an IB switch with FIFO input buffers.

In addition to this technique, the HOL blocking problem for FIFO buffers can be overcome in CICB switches by using virtual output queues (VOQs), where a VOQ is a queue in the input that stores cells destined for a specific output. CICB switches with VOQ are denoted as VOQ–CICB (see Figure 6.5). However, for the sake of brevity, we refer to VOQ–CICB switches as CICB switches in the remainder of this chapter.

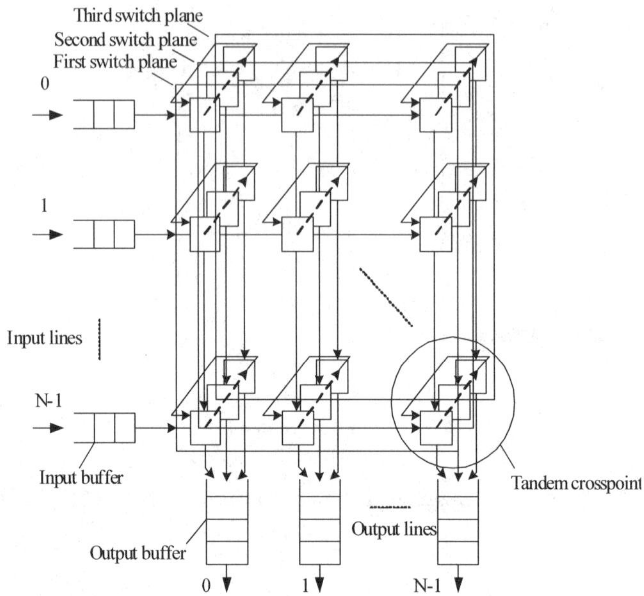

Figure 6.4. Tandem-crosspoint (TDXP) switch with three planes

6.4.2 VOQ–CICB Switches

Consider a CICB switch with N inputs and outputs. In this switch model, there are N VOQs at each input. A VOQ at input i, where $0 \leq i \leq N-1$, that stores cells for output j, where $0 \leq j \leq N-1$, is denoted as $VOQ_{i,j}$. A crosspoint (CP) element in the CICB that connects input port i to output port j is denoted as $CP_{i,j}$. The buffer at $CP_{i,j}$ is denoted as $CPB_{i,j}$. The size of $CPB_{i,j}$, k, is indicated by the number of cells that can be stored. A credit-based flow-control mechanism indicates to input i whether $CPB_{i,j}$ has room available for a cell or not, as described in [15]. $VOQ_{i,j}$ is said to be eligible for selection if the VOQ is not empty and the corresponding $CPB_{i,j}$, at BC, has room to store a cell.

The round-trip (RT) time, as in [15], is defined as the sum of the delays of the input arbitration (IA), the transmission of a cell from an input to the crossbar ($d1$), the output arbitration (OA), and the transmission of the flow-control information back from the crossbar to the input ($d2$). Figure 6.6 shows an example of RT for input 0 by showing the transmission delays for $d1$ and $d2$, and arbitration times, IA and OA. Cell and bit alignments are included in the transmission times. The following condition is for this switch to avoid underflow

$$RT = d1 + OA + d2 + IA \leq k \tag{6.1}$$

where k is the crosspoint buffer size (in time slots) which is equivalent to the number of cells that can be stored. In other words, the crosspoint buffer must be able to store

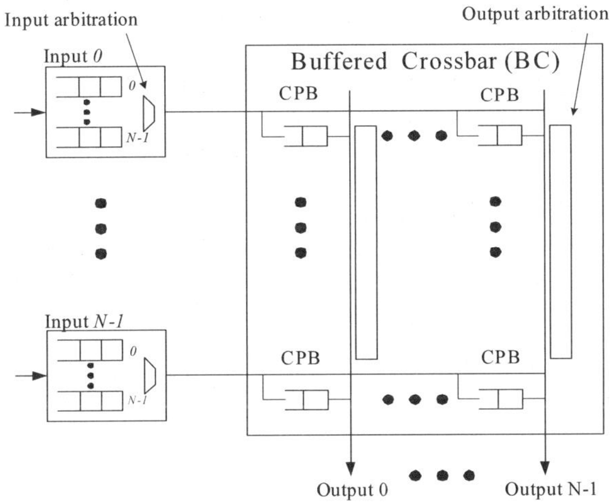

Figure 6.5. $N \times N$ buffered crossbar with VOQs

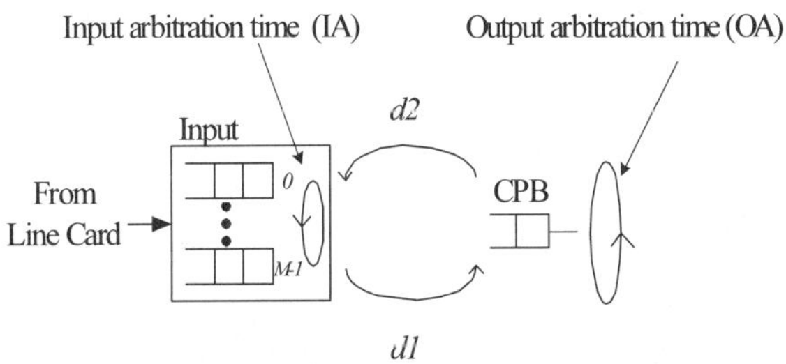

d1: Transmission delay from input *i* to crossbar

d2: Transmission delay from crossbar to input *i*

Round Trip Time = IA + *d1* + OA + *d2*

Figure 6.6. Demonstration on the credit-base flow control

a number of cells to keep the buffer busy (*i.e.* transmitting cells) during at least one RT time.

As the buffered crossbar switch can be physically located far from the input ports, actual round trip times can be non-negligible. To support non-negligible round-trip time in a buffered-crossbar switch, the crosspoint buffer size needs to be increased, such that up to RT cells can be buffered. Non-negligible round-trip delays have been considered recently in [16], [17], and [18], for practical implementations.

6.4.3 Separating Matching into Input and Output Arbitrations

CICB switches use selection time efficiently as input and output port selections are performed separately.[2] For each input in a CICB switch, there is one input arbiter, which separately resolves input contention, *i.e.* a VOQ is selected to transfer a cell into the switch core. The arbitration of an input is independent from the arbitration of the other inputs, and from output arbitration. In a similar way, for each output, there is one output arbiter which independently deals with output contention, *i.e.* decides which crosspoint buffer is allowed to transfer a cell out of the switch core. The output arbitration is separated from the arbitration of other outputs and from input arbitration. The input and output arbiters are only coupled by the flow control mechanism [19]. Each crosspoint buffer and corresponding VOQ has an associated credit which is used as a flag for the state of the crosspoint buffer (1=full, 0=empty). During the input scheduling phase, the input scheduler at each input i selects a non-empty $VOQ_{i,j}$ whose credit state is 0 and sets the credit status to 1. During the output scheduling phase, the output scheduler at each input j selects a non-empty crosspoint buffer (i, j) whose credit state is 1 and sets the credit status to 0. Back to the example of the stringent timing, a CICB switch with 40-Gbps and 80-byte packets can perform input (or output) arbitration within 16 ns, and therefore, the timing for arbitration is extended.

The use of VOQs in a CICB switch allows us to apply different selection (or arbitration) schemes at the inputs, and the existence of crosspoint buffers at an output allows different selection schemes to be selected at the outputs. We can divide the arbitration schemes in CICB switches into weight-based schemes and weightless schemes. Weight-based schemes assign a weight to each contending queue and the arbiter selects the one with the largest weight. The weight of a queue is assigned according to the measurement of a common parameter in all contending queues, such as the age of the cells. A weightless scheme assigns the same weight (usually 1) to all queues, and queue selection is performed according the the order they were selected in the past or any arbitrary rule (*e.g.* random). Input and output arbitration can use the same or different schemes. They can also be a combination of weighted and weightless selection schemes. The following sections present several combinations of input and output arbitration schemes in a CICB switch.

6.4.4 Weighted Arbitration Schemes

Oldest-cell First (OCF) and Round-robin Arbitrations

One of the advantages of having large input buffers in a switch is that weights can be assigned to VOQs in different ways. Therefore, an input arbiter can perform VOQ selection in a weight-based fashion. A CICB with weight-based input arbitration was presented in [20], where oldest cell first (OCF) selection for input arbitration

[2] Other switches, such as IB switches, need to perform matching; output selection (input arbitration) can be performed after input selection (output arbitration) in a sequential manner.

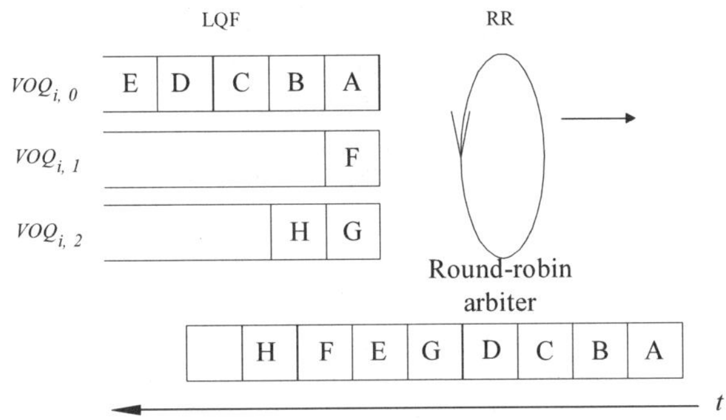

Figure 6.7. Example of LQF-RR among three queues in a 3 × 3 switch

and round-robin selection for output arbitration were used. We denote this combination as OCF-RR. A CICB switch using this combination showed high performance compared to an IB switch.

This CICB switch with timestamp-based arbitration and VOQs at the input ports showed that the crosspoint-buffer size can be small if the VOQs are provided with enough storage capacity. This approach was also studied in [23].

Longest Queue First and Round-robin Arbitrations

One of the advantages of using weight-based arbitrations is that stability can be studied. Stability of a switch provides information about the maximum throughput, *i.e.* 100%, by a switch. In general, a switch is considered stable if the occupancy of the input buffers is stable. This means that the occupancy size is bounded. The longest queue first (LQF) selection [21] and the longest normalized queue first (LNQF) selection [22] have been considered and analyzed for IB switches. With that, LQF was applied into CICB switches [23]. Through a fluid model [24], it was proved that a CICB switch with the combination of LQF as input selection and RR as output selection (LQF-RR), with one-cell crosspoint buffers, obtains 100% throughput, where each input–output pair has a load no more than $1/N$ (*i.e.* traffic with a uniform distribution).

Most Critical Internal Buffer First

One of the reasons LQF was not applied to output arbitration schemes is because switches were considered with one-cell crosspoint buffers, where the weights could only be $\{0,1\}$, making LQF ineffective at the outputs. An alternative for weight assignment is presented in [25]. Here, a scheme named Most Critical Internal Buffer First (MCBF) is proposed. This scheme does not consider the state of a single crosspoint buffer, but rather the set of crosspoint buffers in the buffered crossbar. MCBF favors the least occupied internal buffer at the input side, named shortest internal buffer first (SBF), while the output favors the most occupied internal buffer, named longest internal buffer first (LBF). MCBF is based only on the crosspoint buffer information, by which the MCBF scheme makes good use of the CICB switch's advantage provided by the crosspoint buffers.

The line of crosspoint buffers $LCPB_i$ is defined as the set of all the internal buffers ($CPB_{i,j}$) that correspond to the same input i and holding cells for all the outputs. NLB_i is the number of cells in $LCPB_i$. The column of crosspoint buffers $CCPB_j$ is defined as the set of the internal buffers ($CBP_{i,j}$) that correspond to the same output j and receiving cells for all the inputs. NCB_j is the number of cells in $CCPB_j$. The MCBF scheme is based on the shortest internal buffer first (SBF) as input arbitration and on the longest internal buffer first (LBF) for output arbitration. These arbitration schemes work as follows:

- Input arbitration SBF scheme: for each input i, select the first eligible VOQ corresponding to $min_j \ NCB_j$ and send its HOL cell to the internal buffer $CPB_{i,j}$.
- Output arbitration LBF scheme: for each output j, select the first $XP_{i,j}$ corresponding to $max_i \ NCB_i$ and send its HOL cell to the output.

The MCBF scheme has three major properties. First, MCBF has simple hardware complexity when compared to LQF-RR and OCF-OCF, because MCBF's arbitration decisions are based on the number of cells in the internal buffers (NLB_i, NCB_j). For a $N \times N$ CICB switch with one-cell crosspoint buffer, an arbiter's encoder consists only of log N bits. Second, MCBF is a scheme which is almost stateless. It performs arbitration without any type of state information about the input VOQs. The only feedback information that MCBF needs to have during its arbitration process is whether an input VOQ is empty or not. Finally, MCBF is designed to be a matched pair of input and output in finding matched scheduling. No output is idle so long as $NCB_j \geq 1$.

Figure 6.8 shows an example of VOQ selection by an input arbiter in a 4×4 switch. Figure 6.8(a) shows the state of the CPBs with 1 for those that are active and 0 for those CPBs that are idle (empty). Figure 6.8(b) shows the column occupancy as seen by each input (matrix row). For example, input 1, the second row, has two CPBs available (*i.e.* idle), $CPB_{0,2}$ and $CPB_{0,3}$ (assume that $0 \leq i, j \leq 3$, where input 0 is the top row and output 0 is the left-most column). Therefore, input 0 can choose to serve a cell from $VOQ_{0,2}$ or $VOQ_{0,3}$. The column occupancy is then 2 and 1 cells for output 2 and 3, respectively. Therefore, using SBF, $VOQ_{0,3}$ is selected.

Figure 6.9 shows an example of the selection process of a CPB that LBF performs in a 4×4 buffered crossbar, which is represented as a matrix. In this matrix, rows represent inputs and columns represent outputs. A CPB with a cell is represented by 1 (busy), and 0 (idle), otherwise. Figure 6.9(a) shows the state of CPBs as busy and idle crosspoints. Figure 6.9(b) shows the input occupancy seen by the output arbiters per CPB. For example, $CPB_{0,0}$ is 1 as there is only one cell from input 0 for all outputs. An Idle CPB is indicated by a zero as it is ignored by the output arbiter. This figure also shows that this example has all CPBs from inputs 1 with the longest input occupancy, and the output arbiters, using the same selection policy, select all CPBs

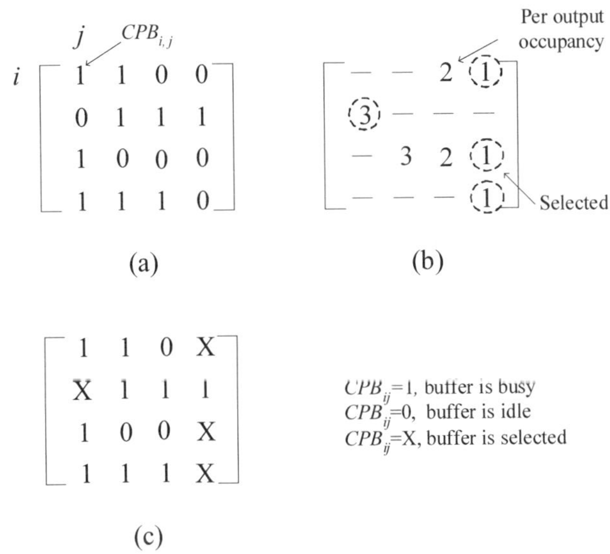

Figure 6.8. Example of VOQ selection by an input arbiter using SBF in a 4×4 switch

from input 1. Therefore, Figure 6.9(c) shows that $CPB_{1,1}$ to $CPB_{1,3}$ are selected and also $CPB_{2,0}$ is selected, marked with an X.

In a similar way, another scheme named CBF was presented in [26]. CBF differs from MCBF in the measurement of criticalness of the internal buffer. For MCBF, it is in terms of the queue length of the internal buffer, while for CBF, it is in terms of the HOL cell age of the internal buffer.

Performance of Weight-based Arbitration Schemes

The addition of crosspoint buffers in a switch improves the switching performance when compared to a switch with no crosspoint buffers (*e.g.* IB switches). In this section, we present simulation results of weight-based arbitration schemes under admissible traffic. Here, we consider admissible traffic defined as follows. Denote $\lambda_{i,j}$ as the cell arrival rate at input i for output j that is received in $VOQ_{i,j}$. The traffic is considered admissible if

$$\sum_i \lambda_{i,j} \leq 1 \tag{6.2}$$

and

$$\sum_j \lambda_{i,j} \leq 1 \tag{6.3}$$

CICB switches with these weight-based schemes have been simulated under uniform traffic with Bernuolli arrivals. In these simulation, it has been observed that the

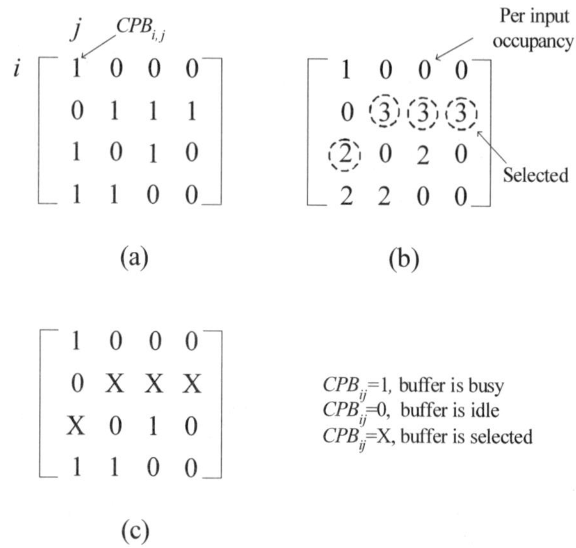

Figure 6.9. An example of CPB selection by an output arbiter using LBF in a 4×4 switch

average cell delay (queuing delay) is close to that of an OB switch. However, under non-inform traffic, these schemes provide different performance results.

One of the traffic patterns used to test this is the unbalanced traffic model [15]. The unbalanced traffic model uses a probability w as the fraction of input load directed to a single predetermined output, while the rest of the input load is directed to all outputs with uniform distribution. Consider input port s, output port d, and the offered input load for each input port ρ. The traffic load from input port s to output port d, $\rho_{s,d}$, is given by

$$\rho_{s,d} = \begin{cases} \rho\left(w + \frac{1-w}{N}\right) & \text{if } s = d \\ \rho\frac{1-w}{N} & \text{otherwise} \end{cases} \tag{6.4}$$

When $w = 0$, the offered traffic is uniform. On the other hand, when $w = 1$, it is completely directional, from input s to output d, where $s = d$.

Figure 6.10 shows the performance of OCF, LQF, and MCBF arbitration schemes under unbalanced traffic for switches with a crosspoint buffer with size equal to one cell. This figure shows that LQF and OCF deliver close to 100% throughput for all values of w. MBCF (labeled SBF-LBF) delivers below 99% throughput for some values of w; however, the throughput is higher than that delivered by an IB switch. As MCBF depends on the occupancy of the crosspoint buffers, it is expected that the performance improves as the size of the crosspoint buffer increases.

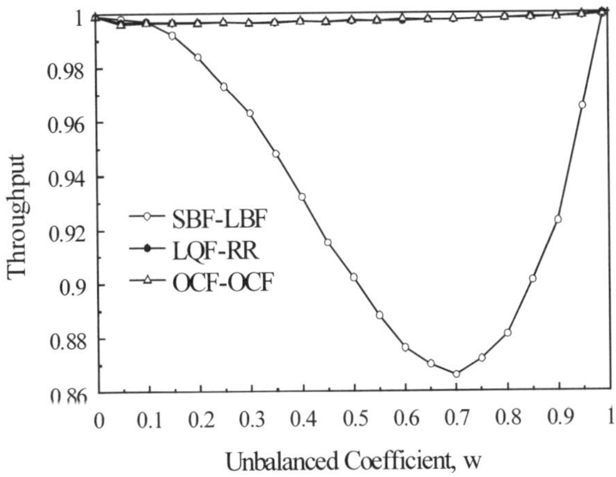

Figure 6.10. Performance of weight-based arbitration schemes

6.4.5 Arbitration Schemes based on Round-robin Selection

Weight-based schemes need to perform comparisons among all contending queues (or crosspoint buffers), which can be a large number, and this may increase the implementation complexity. Moreover, weight-based schemes (*e.g.* queue-occupancy based) may starve some queues for a long time to provide more service to congested queues, thus presenting unfairness. On the other hand, RR algorithms have been shown to provide fairness and implementation simplicity (as no comparisons are needed among queues) and high performance under uniform traffic [27]. However, schemes based on round-robin selection have not been shown to provide near 100% throughput under nonuniform traffic patterns with a buffered crossbar having small size crosspoint buffers. It has been shown that a switch using RR needs a large crosspoint buffer to provide high throughput under admissible unbalanced traffic [28], where the unbalanced traffic model is a nonuniform traffic pattern [15]. This large buffer can make the implementation of a switch costly. Here, we review some of the round-robin schemes for delivering high performance.

Round-robin (RR) Arbitration

Round-robin (RR) selection has been used in IB switches to provide high throughput under uniform traffic [29]. RR is attractive because of the high degree of service fairness and its simple implementation, necessary for high-speed switches. RR has been used in CICB switches [15], [23]. A switch using RR, called CIXB-1, has been used to show that a CICB switch using one-cell crosspoint buffers and credit-based flow control can provide 100% throughput for uniform traffic [15]. In CIXB-1, the input arbiter selects a non-inhibited and non-empty VOQ. The inhibition of VOQs is determined by the flow control mechanism. A VOQ is said to be inhibited if the crosspoint

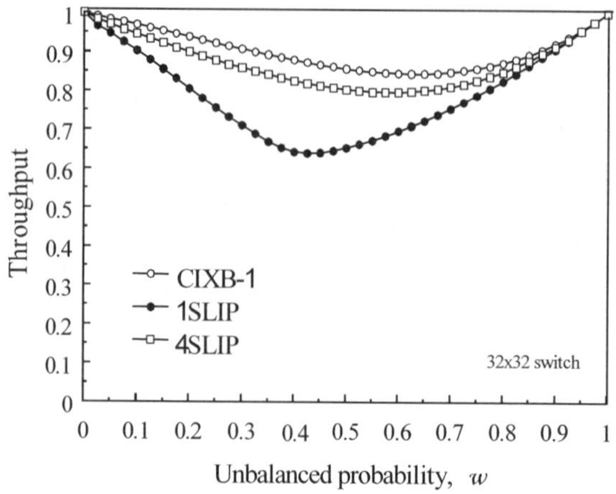

Figure 6.11. Performance of CIXB-1, *i*SLIP under unbalanced traffic

buffer has no room for a new cell. If an input sends a cell from an inhibited VOQ, then the cell would be lost. The output arbiter performs selection of a crosspoint in a round-robin fashion, where only non-empty crosspoint buffers are considered. Since there are no buffers in the output, an output arbiter is not needed to check for inhibition. After a RR arbiter selects a VOQ (for the input) or a crosspoint buffer (for the output), the pointer is moved to one position beyond the selected VOQ.

The CIXB-1 switch has some attractive properties: a simple arbitration scheme and relaxed timing for arbitration. The arbitration complexity is $O(N)$, which can be reduced to $O(\log N)$ by suitable encoding logic. This switch has been shown to provide 100% throughput under uniform traffic with Bernoulli and bursty arrivals. CIXB-1 was also used to show that CICB switches can deliver higher throughput than IB switches, as presented in a comparison of CIXB-1, *i*SLIP, and an OB switch.

The throughput of CIXB-1 and *i*SLIP for unbalanced traffic is shown in Figure 6.11. The performance of CIXB-1 is higher than 1-SLIP and 4-SLIP. The throughput of CIXB-1 is almost 100% when w is about zero (*i.e.* uniform traffic) and about one (*i.e.* totally directional traffic).

Compared with the *i*SLIP algorithm, CIXB-1 has better delay performance under various traffic patterns. However, as actual traffic may present nonuniform distributions, it is necessary to provide arbitration schemes that provide 100% throughput for admissible traffic.

Round-robin with Adaptable Frame Size (RR-AF) Arbitration Scheme

As discussed in the previous section, weight-based arbitration schemes, where weights are assigned to input queues proportionally to their occupancy or HOL cell age, can provide high throughput under a wide variety of admissible traffic patterns [23], [30]. However, weight-based schemes need to perform comparisons among all contending queues, which can be a large number, thus having high implementation complexity. Moreover, weight-based schemes (*e.g.* queue-occupancy based) may starve some queues for long time to provide more service to congested ones, presenting a degree of unfairness. On the other hand, round-robin schemes have been shown to provide fairness and implementation simplicity, as no comparisons are needed among queues. Furthermore, these schemes are known to have high-performance under uniform traffic [27]. However, schemes based on round-robin selection have not been shown to provide near 100% throughput under nonuniform traffic patterns with a buffered crossbar having crosspoint buffers of small size. For example, it has been shown that a switch using RR needs a large crosspoint buffer to provide high throughput under admissible unbalanced traffic [28], where the unbalanced traffic model is a nonuniform traffic pattern [15]. This large buffer can make the implementation of a switch costly.

Frame-based matching has been shown to have improved switching performance under different traffic scenarios [31]. However, how to set the frame size is a complex issue. The round-robin with adaptable-size frame (RR-AF) scheme was proposed to avoid assigning frame sizes arbitrarily [32]. RR-AF is based on the amount of service that a buffers gets. Each time a VOQ (or a CPB at an output) is selected by the arbiter, the VOQ gets the right to forward a frame, where a frame is formed by one or more cells. Each cell of a frame is dispatched in one time slot. The RR-AF scheme can achieve near 100% throughput under admissible uniform and nonuniform traffic.

In RR-AF, each VOQ (and CPB) has two counters: a frame-size counter, $FSC_{i,j}(t)$, and a current service counter, $CSC_{i,j}(t)$. The value of $FSC_{i,j}(t)$, $|FSC_{i,j}(t)|$, indicates the frame size; that is, the maximum number of cells that $VOQ_{i,j}$ can send in back-to-back time slots to the buffered crossbar, one cell per time slot. The initial value of $|FSC_{i,j}(t)|$ is one cell (*i.e.* its minimum value). It is considered that $|FSC_{i,j}(t)|$ can be as large as needed, although practical results have shown that its value is not large. $CSC_{i,j}(t)$ counts the number of serviced cells at time slot t in a frame corresponding to a VOQ, where the frame size is indicated by FSC, in a regressive fashion. A regressive-fashion count is used in CSC as CSC only considers FSC at the end of a serviced frame. The initial value of $CSC_{i,j}(t)$, $|CSC_{i,j}(t)|$, is one cell (*i.e.* its minimum value).

The input arbitration process works as follows. An input arbiter selects an eligible $VOQ_{i,j'}$ in round-robin fashion, starting from the pointer position j. For the selected $VOQ_{i,j'}$, if $|CSC_{i,j'}(t)| > 1$, $|CSC_{i,j'}(t+1)| = |CSC_{i,j'}(t)| - 1$, and the input pointer remains at $VOQ_{i,j'}$, so that this VOQ has the higher priority for service in the next time slot and the frame transmission can continue. If $|CSC_{i,j'}(t)| = 1$, the input pointer is updated to $(j'+1)\text{mod}(N)$, $|FSC_{i,j'}(t)|$ is increased by f cells, and $|CSC_{i,j'}(t)| = |FSC_{i,j'}(t)|$. For any other $VOQ_{i,h}$, where $h \neq j'$, which is empty or inhibited by the flow-control mechanism, and it is positioned between the pointed $VOQ_{i,j}$ and the selected $VOQ_{i,j'}$: if $|FSC_{i,h}(t)| > 1$, $|FSC_{i,h}(t+1)| = |FSC_{i,h}(t)| - 1$. If there exist one or more VOQs that fit the description of $VOQ_{i,h}$ at a given time slot, it is said that those VOQs missed a service opportunity at that time slot. The increment of the frame size, done by f cells, is performed each time the previous complete frame of a VOQ has been serviced. For the sake of clarity, the following pseudo-code describes the input arbitration scheme, as seen at an input:

-At time slot t, starting from the pointer position j*, find the nearest eligible* $VOQ_{i,j'}$ *in a round-robin fashion.*
-Send the HOL cell from $VOQ_{i,j'}$ *to* $XPB_{i,j'}$ *time slot* $t+1$.
If $|CSC_{i,j'}(t)| > 1$ *then*
$|CSC_{i,j'}(t+1)| = |CSC_{i,j'}(t)| - 1$,
the pointer points to j'.
else $|FSC_{i,j'}(t+1)| = |FSC_{i,j'}(t)| + f$,
$|CSC_{i,j'}(t+1)| = |FSC_{i,j'}(t+1)|$,
the pointer points to (j'+1) module N.
-For VOQ(i, h), where $j \le h < j'$ *for* $j < j'$*, or* $0 \le h < j'$ *and* $j \le h \le N-1$ *for* $j > j'$*:*
$FSC_{i,h}(t+1) = FSC_{i,h}(t) - 1$.[3]
-Go to the next time slot.

Note that f may be equal to a constant or a variable value. In general, f assumes the finite value of N, unless otherwise stated. The value of f affects the performance of RR-AF in different traffic scenarios. Note that when $f = 0$, RR-AF becomes RR.

Figure 6.12 shows an example of the adjustment of $FSC_{i,j}$ in an input of a 4×4 switch. In this example, $VOQ_{i,2}$ and $VOQ_{i,3}$ have cells (as Figure 6.12(a) shows), one and three, respectively, and no VOQ is inhibited by the flow-control mechanism. At time slot t, the pointer of RR-AF points to $VOQ_{i,0}$. During this time slot, the input arbiter selects $VOQ_{i,2}$ to send a cell to the buffered crossbar. Then, $VOQ_{i,0}$ and $VOQ_{i,1}$ miss an opportunity to send cells as they are empty and their FSCs are decreased by one at the end of the time slot. Note that $VOQ_{i,0}$ and $VOQ_{i,1}$ are considered $VOQ_{i,h}$ for this time slot as defined in the description of RR-AF. Table 8.1 shows the evolution of the FSC values for each VOQ during 6 time slots. In the next time slot, $t+1$, $VOQ_{i,2}$ is served, and it becomes empty. As the pointer points to this VOQ, $FSC_{i,2}$ is decreased to 1 in the next time slot. Therefore, the arbiter selects $VOQ_{i,3}$ at time slot $t+2$ as the next VOQ to receive service. Then, the pointer is moved to $VOQ_{i,3}$. At time slot $t+3$, $VOQ_{i,3}$ is again selected. Since the last frame cell of $VOQ_{i,3}$ is selected, $FSC_{i,3}$ is updated to $2 + N = 2 + 4 = 6$. However, since there are no more cells in this VOQ, $FSC_{i,3}$ decreases by one in the subsequent time slots. In this table, a dash in time slots $t+4$ and $t+5$ means that no j is selected. Figure 6.12(b) shows the order in which cells are served.

RR-AF delivers 100% throughput under uniform traffic. It also has an average cell delay close to that of an OB switch, as RR selection does. Furthermore, RR-AF shows no measurable improvement under uniform traffic with Bernoulli arrivals and with different CPB sizes.

However, the efficiency of RR-AF can be better appreciated under unbalanced traffic, as RR-AF, with a CPB size of one cell can provide better performance (higher

[3] Note that when $j' = j$, there is no $VOQ(i, h)$.

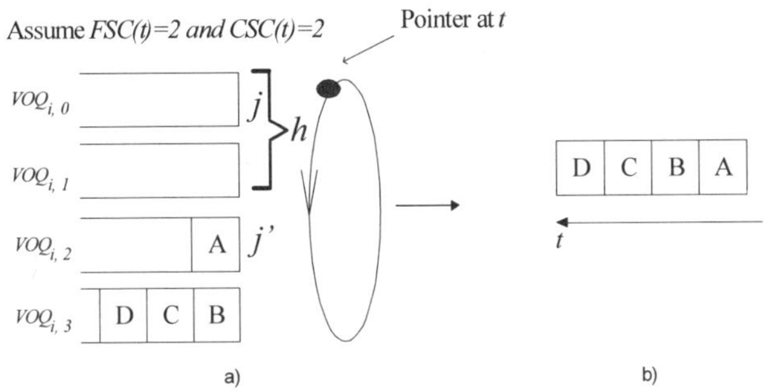

Figure 6.12. Example of VOQs and their FSC values in a 4×4 switch with RR-AF

Table 6.1. Evolution of FSC of example in Figure 6.12

	Time slot					
FSC	t	$t+1$	$t+2$	$t+3$	$t+4$	$t+5$
$FSC_{i,0}$	2	1	1	1	1	1
$FSC_{i,1}$	2	1	1	1	1	1
$FSC_{i,2}$	2	2	1	1	1	1
$FSC_{i,3}$	2	2	2	6	5	4
Selected j	2	3	3	3	-	-

throughput) than RR with a CPB size of 32 cells in a 32×32 switch. Figure 6.13 shows the simulation results of RR-AF.

Round-robin with Frame-size Occupancy-based (RR-FO) Arbitration

The round-robin with frame-size occupancy-based arbitration (RR-FO) scheme was proposed in [33]. The scheme is also round-robin based. A new frame is assigned a size, in number of cells, each time the previous frame is serviced. The frame size is determined by the cell occupancy of the VOQ at the time the frame service completes. This is called captured-frame size. A VOQ is said to be in on-service status if the VOQ has a frame size of two or more cells and the first cell of the frame has been transmitted. An input is said to be on-service if there is at least one on-service VOQ. A VOQ is said to be off-service if the last cell of the VOQ's frame has been sent, or no cell of the frame has been sent to the buffered crossbar. At time t_c of selecting the last cell of a frame of $VOQ(i,j)$, the next frame is assigned a size equal to the occupancy of $VOQ(i,j)$. Cells arriving at $VOQ(i,j)$ at any time t_d, where $t_d > t_c$, are considered for selection until the current frame is totally served and they are included in a new captured frame.

For each VOQ, there is a captured frame-size counter, $CF_{i,j}(t)$. The value of $CF_{i,j}(t)$, $|CF_{i,j}(t)|$ indicates the frame size, that is, the maximum number of cells that a $VOQ(i,j)$ has as candidates in the current and future time slot, one per time

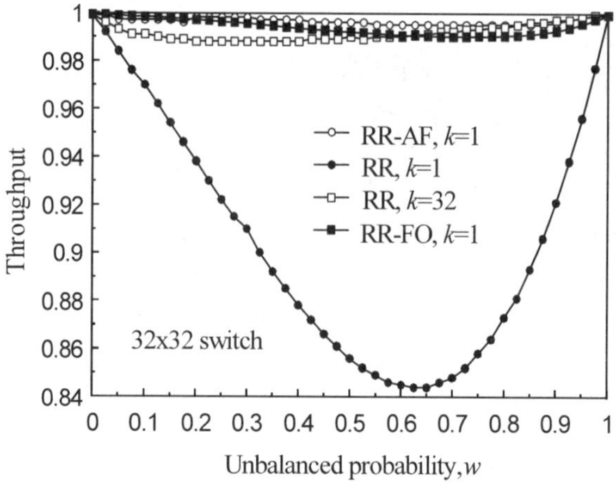

Figure 6.13. Throughput performance of round-robin-based schemes under unbalanced traffic

slot. $|CF_{i,j}(t)|$ takes a new value when the last cell of the current frame of $VOQ(i,j)$ is selected. $|CF_{i,j}(t)|$ decreases its count each time a cell is selcted, other than the last.

The input arbitration process works as follows: input arbiter selects an eligible on-service VOQ in round-robin order, starting from its pointer position. If no on-service VOQ is present, an off-service VOQ is selected in round-robin order.

- If the input arbiter selects $VOQ(i,j)$ and if $|CF_{i,j}(t)| > 1$: $|CF_{i,j}(t+1)|$=$|CF_{i,j}(t)|$-1 and this VOQ is set as on-service.
- Else (*i.e.* $|CF_{i,j}(t)|$=1): $|CF_{i,j}(t+1)|$ is assigned the occupancy of $VOQ(i,j)$, and $VOQ(i,j)$ is set as off-service. The pointer at the input arbiter then moves to one position beyond the accepted $VOQ(i,j)$: to the next output or $(j+1)$ module N.

The output arbitration process is the round-robin selection with pointer persistency. After $CPB(i,j)$ has been selected by the output arbiter, the output pointer keeps pointing to input i.

RR-FO, as RR and RR-AF, provides 100% throughput and an average cell delay close to that of an OB switch under uniform traffic. Under unbalanced traffic, RR-FO gives similar performance to that of RR-AF. Figure 6.13 shows the throughput of RR-AF, RR-FO, and RR under unbalanced traffic.

The switch with RR-AF and RR-FO uses a CPB size of one cell, and the RR uses a CPB size of one and 32 cells, respectively. As an example, RR, with 32-cell CPBs and a cell size of 64 bytes, would need 16 Mb of memory, while RR-AF or RR-FO, with 1-cell CPBs, would need 512 Kb of memory. In addition, RR-AF and RR-FO provide higher performance than RR.

6.5 CICB Switches with Internal Variable-length Packets

The switches discussed so far perform internal transmission of fixed-size packets, called cells. In this section, we discuss a different approach: the switching of variable-size packets. Therefore, switches for variable length packets do not perform segmentation and re-assembly of packets.

Using cells internally in a switch requires that padding bytes be transferred when a packet length is not evenly divisible by the cell length. In the worst case, there could be cells with a size of S bytes and an incoming packet with size of $S+1$ bytes. To transmit this packet, two time slots would be needed. Therefore, to maintain wirespeed transmission, an internal speedup of $2S/S+1$, which is approximately 2, is needed [34], if no sophisticated segmentation is used.

A segmentation mechanism in which each segment contains data from different packets, is considered in [35]. In this way, padding bytes are no longer needed and a CICB can keep up with wirespeed transmission without the need for speedup.

A CICB switch using parallel polling was used to switch variable-length packets [36]. This switch used a round-robin order for polling VOQs and CPBs at the inputs and outputs, respectively. The performance of this switch was compared to that of iSLIP in an IB switch. The results also showed that CICB switches have higher performance (lower average packet delay) than IB switches.

Another example of a CICB for variable-length packets was presented in [37], where the implementation of such a switch is discussed in detail. This switch also uses round-robin selection for input and output arbitration.

6.6 Output Emulation by CICB Switches

The OB switch is often considered as the ideal packet switching architecture for providing quality-of-service (QoS) guarantees. If a switch can exactly emulate an OB switch, it means the switch can provide the same throughput as an OB switch. OB switch emulation often requires a shadow OB switch to determine the correct departure time of every cell in the switch. A shadow OB switch is defined as a theoretical OB switch that determines the departure order and time of each cell from a buffered crossbar to exactly emulate an OB. OB emulation implies the use of some other type of switch such as IB, CIOB, and CICB to emulate an OB in order to evaluate performance.

In [38], it was shown that a speedup of 4 is sufficient for a CIOB to exactly emulate an OB switch with the scheduling policy "most urgent cell first" (MUCF). However, with a speedup of 2, a CIOB behaves identically like an OB, and a speedup of $2-\frac{1}{N}$ with a scheme named "critical cell first" (CCF) is sufficient to mimic a FIFO-OB switch [39].

In [40] and [41], it was shown that a CICB switch with a speedup of 2 can mimic a FIFO-OB switch with any arrival traffic pattern. This is called combined input-crosspoint-output buffered (CICOB) [28]. In [39], it was proven that with a weighted round-robin scheduler, a CICOB can achieve 100% throughput and can mimic an OB

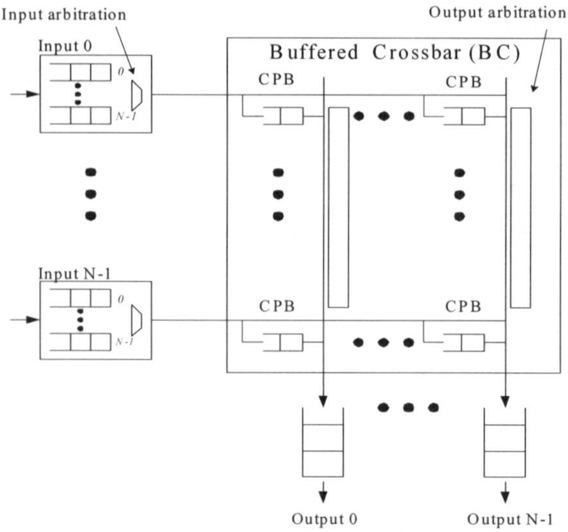

Figure 6.14. $N \times N$ buffered crossbar with output queues

switch with a speedup of 2 for admissible traffic. Note that [39] and [40] require that the buffered crossbar switches work with a speedup of 2, where the switches have output queues.

The following terms are defined to sketch the proof:

Time to leave TTL(*c*) is the time slot during which cell c departs as specified by the shadow OQ switch.

Input priority list (IPL): each input scheduler maintains an input priority list of all cells queued at input i. The IPL of an input scheduler determines the departure order of cells from the input to the internal buffers.

Output priority list (OPL): each output scheduler maintains an output priority list of all cells queued at output j.

Input thread IT(*c*) is the number of cells ahead of c in its input priority list; IT(c) is defined for each cell queued at an input port. IT(c) may decrease by one during a scheduling phase and may increase by one during the arrival phase. When a cell is transferred from the input queue to the crosspoint, IT(c) is set to zero.

Output cushion OC(*c*) is the number of cells at c's output queue with lower TTL than c. OC(c) may increase by one during a scheduling phase and decrease by one during the departure.

Slackness equals the output cushion of cell c minus its input thread, which is L(c)=OC(c)−IT(c). The slackness reflects the urgency with which the switch must transfer the cell to its output. These terms are exemplified in a CICOB switch, as show in Figure 6.15.

The OB emulation process is highly influenced by the slackness of every cell c inside the system. Any increase in L(c) is translated by either an increase in OC(c)

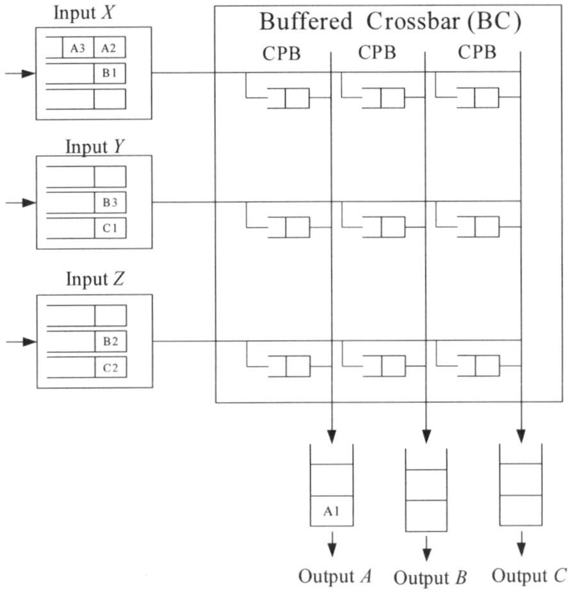

IT(A3)=1, OC(A3)=2, L(A3)=1

Figure 6.15. Example of IT and OC in a 3 × 3 CICOB switch

or decrease in IT(c). In both cases, L(c) increases, and thus c will reach its output on time. Any decrease in L(c) is translated by either a decrease in OC(c) or an increase in IT(c). In both cases, L(c) decreases, and c should be urgently transferred to its output queue before it misses its time to leave. As a result, in order for OB emulation to occur, the slackness of every cell inside the switch is always positive and non-decreasing.

A CICOB switch with a speedup of 2 can exactly emulate an OB switch. With a speedup of 2, each time slot is divided into the following four phases.

1. Arrival phase: all arriving cells are received and inserted in an input queue.
2. First scheduling phase selects and transfers cells from the crosspoint queue to the output queues, and from input queues to the crosspoint queues.
3. Second scheduling phase selects and transfers cells from the crosspoint queue to the output queues, and from input queues to the crosspoint queues.
4. Departure phase: all departures from the output queues occur in this phase.

Each scheduling phase contains two operations:

(a) Input scheduling: the input scheduling algorithm selects the highest priority cell whose associated flow-control state is 0 from each input's IPL and transfers the cell to the crosspoint buffer.

(b) Output scheduling: the output scheduling algorithm selects the highest priority cell according to the output's OPL and transfers the cell to the output.

At each time slot, every cell c can be in one of the following states: just arrived; selected for input scheduling; not selected for input scheduling (blocked by a flow control); selected for output scheduling; selected for output scheduling (blocked by a more urgent cell) or depart the switch. The OB emulation occurs if, irrespective of its status, any cell c has non-negative slackness.

For any new arriving cell, the slackness is proved to be non-negative. This property is called non-negative slackness insertion. If cell c departs to the output queue in the next time slot, we are no longer concerned with L(c). If cell c remains, L(c) increases by at least one in each scheduling phase. During the departure phase, OC(c) decreases by at most one and IT(c) remains unchanged, implying that L(c) may decrease by 1. During the arrival phase, IT(c) increases by at most 1, and OC(c) remains unchanged, implying that L(c) may decrease by one. Summing over the four phases, L(c) does not decrease from time slot t to time slot t+1.

In a FIFO OB switch, a newly arriving cell always has a higher TTL than all cells previously arrived from the same VOQ, and the TTL assigned to the arriving cell remains fixed until the cell departs. In order to emulate a FIFO-OB switch, [40] and [41] proposed different input/output scheduling schemes such as $GBVOQ - OCF$ and $MCAF - LTF$, respectively. By the input insertion algorithm named group-by-virtual-output-queue (GBVOQ) upon arrival, if the input VOQ is empty, the cell is placed at the front of IPL; otherwise, the cell is placed in an IPL position just behind the last cell for the VOQ in the IPL. The OPL for each output is ordered by OCF. In fact, both of them attempt to prove a CICOB switch with their scheduling scheme and a speedup of 2 satisfies the NNS insertion property and a non-decreasing slackness from a time slot to the next; therefore, they can exactly emulate a FIFO-OB switch.

Theoretical analysis has shown that it is possible to emulate an OB switch using a CICOB switch with a speedup of 2 and a suitable scheduling scheme. However, the implementation complexity of the proposed schemes is still too high to be practical.

6.7 Conclusions

This chapter presents an overview of internally buffered crossbar switches as a feasible and attractive alternative to high-performance packet switches for the next generation networks. This type of switch requires that memory works at the same speed as that for IB switches to provide high switching performance. Most internally buffered switches have been shown to outperform IB switches and to emulate OB switches with a speedup no more than 2. Therefore, internally buffered switches result in an economic and efficient solution to the implementation of packet switches. Issues remaining for future research include the provisioning of guaranteed services using internally buffered crossbar switches and the resolution of scalability issues. As line rates continue to increase and memory speedup falls behind, new architectures with more memory-efficient strategies will be needed.

References

1. C. Clos, A study of nonblocking switching networks, *Bell Syst. Tech. J.*, pp. 406-424, Mar 1953.
2. F.A. Tobagi and T Kwok, The Tandem banyan switching fabric: a simple high-performance fast packet switch, Proc. *IEEE Infocom'91*, pp .1245-1253, 1991.
3. M. Karol and M. Hluchyj, Queuing in high-performance packet-switching, *IEEE J. Select. Area Commun.*, **6**, pp. 1587-1597, Dec. 1988.
4. S. Li and N. Ansari, Chapter 1.3: Switch Architecture and Scheduling Algorithms, in *ATM Handbook* (F. Golshani and F. Groom, Eds), International Engineering Consortium, pp. 37-54, 2000.
5. S. Li and N. Ansari, Input-queued switching with QoS guarantees, Algorithms, *Proc. IEEE INFOCOM'99*, March 21-25, 1999, New York, USA, pp. 1152-1159.
6. E.D. Brooks, A butterfly processor–memory interconnection for a vector processing environment, *Parallel Computing* **4**, North Holland, 1987.
7. E. Rathgeb, T. Theimer, and M. Huber, Buffering concepts for ATM switching networks, *IEEE GLOBECOM*, pp. 1277-1281, Dec. 1988.
8. S. Nojima, E. Tsutsui, H. Fukuda, and M. Hashimmoto, Integrated packet network using bus matrix, *IEEE J. Select. Areas Commun.*, **SAC-5**, no. 8, pp. 1284-1291, Oct. 1987.
9. E. Oki and N. Yamanaka, Scalable crosspoint buffering ATM switch architecture using distributed arbitration scheme, *IEEE ATM'97 workshop*, pp. 28-35, 1997.
10. Y. Doi and N. Yamanaka, A high-speed ATM switch with input and cross-point buffers, *IEICE Trans. Commun.*, **E76**, no.3, pp. 310-314, March 1993.
11. D. E. Re and R. Fantacci, Performance evaluation of input and output queuing techniques in ATM switching system, *IEEE Trans. Commun.*, **40**, no. 10, pp. 1565-1575, Oct. 1993.
12. A. K. Gupta, L. O. Barbosa, and N. D. Georganas, 16 x 16 limited intermediate buffer switch module for ATM networks," *GLOBECOM '91*, pp. 939-943, Dec. 1991.
13. A. K. Gupta, L. O. Barbosa, and N. D. Georganas, Limited intermediate buffer switch modules and their interconnection networks for B-ISDN, *ICC '92*, pp. 1646-1650, June 1992.
14. E. Oki and N. Yamanaka, Tandem-crosspoint ATM switch with input and output buffers, *IEEE Communications Letters*, **2**, No. 7, pp. 165-467, July 1998.
15. R. Rojas-Cessa, E. Oki, Z. Jing, and H. J. Chao, CIXB-1: Combined input-one-cell-crosspoint buffered switch, *IEEE HPSR 2001*, pp. 324-329, May 2001.
16. F. Abel, C. Minkenberg,R.P. Luijten, M. Gusat, and I. Iliadis, A four-terabit packet switch supporting long round-trip times, *IEEE Micro*, **23**, Issue 1 , pp. 10-24, Jan.-Feb. 2003.
17. R. Rojas-Cessa, Z. Dong and Z. Guo, Load-balanced combined input-crosspoint buffered packet switch and long-round trip times, *IEEE Communications Letters*, **4**, Issue 7, pp. 661-663, July 2005.
18. Z. Guo and R. Rojas-Cessa, Analysis of a flow control system for a combined input-crosspoint buffered packet switch, *IEEE HPSR 2005*, May 2005.
19. H. T. Kung, K. Chang, Receiver-oriented adaptive buffer allocation in credit-based flow control for ATM networks, *Proc. of IEEE INFOCOM'95*, **1**, pp. 239-252, 1995.
20. M. Nabeshima, Performance evaluation of a combined input- and crosspoint-queued switch, *IEICE Trans. Commun.*, **E83-B**, No. 3, March 2000.
21. N. McKeown, A. Mekkittikul, V. Anantharam, and J. Walrand, Achieving 100% throughput in an input-queued switch, *IEEE Trans. Commun.*, **47**, no. 8, pp. 1260-1267, Aug. 1999.
22. S. Li and N. Ansari, Provisioning QoS features for input-queued ATM switches, *IEE Electronics Letters*, **24**, no. 19, pp. 1826-1827, Sep. 17, 1998.

23. T. Javadi, R. Magill, and T. Hrabik, A high-throughput algorithm for buffered crossbar switch fabric, *IEEE ICC 2001*, pp.1581-1591, June 2001.
24. J. G. Dai and B. Prabhakar, The throughput of data switches with and without speedup, *Proceedings of IEEE INFOCOM 2000*, pp.556-564, March 2000.
25. L. Mhamdi and M. Hamdi, MCBF: a high-performance scheduling algorithm for buffered crossbar switches, *IEEE Commun. Letters*, **7**, Issue 9, pp. 451-453, September 2003.
26. L. Mhamdi and M. Hamdi, CBF: a high-performance scheduling algorithm for buffered brossbar switches, *IEEE HPSR*, pp. 67-72, June 2003.
27. N. McKeown, The *i*SLIP scheduling algorithm for input-queued switches, *IEEE/ACM Trans. Networking*, **7**, no. 2, pp. 188-201, April 1999.
28. R. Rojas-Cessa, E. Oki, and H. J. Chao, CIXOB-k: Combined input-crosspoint-output buffered packet switch, *IEEE GLOBECOM 2001*, **4**, pp. 2654-2660, Nov. 2001.
29. N. McKeown, Scheduling algorithms for input-queued cell switches, Ph.D. dissertation, Dept. Elect. Eng. Comput. Sci., Univ. California at Berkeley, Berkeley, CA, 1995.
30. L. Mhamdi and M. Hamdi, Practical scheduling algorithm for high-performance packet switches, *IEEE ICC 2003*, **3**, pp. 1659-1663, May 2003.
31. A. Bianco, M. Franceschinis, S. Ghisolfi, A.M. Hill, E. Leonardi, F. Neri, and R. Webb, Frame-based matching algorithms for input-queued switches, *IEEE HPSR 2002*, pp. 69-76, May 2002.
32. R. Rojas-Cessa and E. Oki, Round-robin selection with adaptable-size frame in a combined input-crosspoint buffered switch, *IEEE Communications Letters*, **7**, No. 11, pp. 555-557, Nov. 2003.
33. R. Rojas-Cessa, High-performance round-robin arbitration schemes input-crosspoint buffered switches, *IEEE HPSR 2004*, pp. 167-171.
34. D. C. Stephens and H. Zhang, Implementing distributed packet fair queuing in a scalable switch architecture, *Proc. of IEEE INFOCOM'98*, **1**, pp. 282-290, March 1998.
35. M. Katevenis and G. Passas, Variable-size multipacket segments in buffered crossbar (CICQ) architectures, *IEEE ICC*, 2005.
36. K. Yoshigoe and K. J. Christensen, A parallel-polled virtual output queued switch with a buffered crossbar, *Proc. of IEEE HPSR*, pp. 271-275, May 2001.
37. M. Katevenis, G. Passas, D. Simos, I. Papaefstathiou, and N. Chrysos, Variable packet size buffered crossbar (CICQ) switches, Proc. *IEEE ICC 2004*, **2**, pp. 1090-1096, June 2004.
38. B. Prabhakar and N. McKeown, On the speedup required for combined input and output queued switching, *Automatica*, pp. 1909-1920, 1999.
39. S-T. Chuang, S. Iyer and N. McKeown, Practical algorithms for performance guarantees in buffered crossbars, *IEEE INFOCOM*, 2005.
40. R. Magill, C.E. Rohrs and R. L. Stevenson, Output-queued switch emulation by fabrics with limited memory, *IEEE Journal on Selected Areas in Communications*, **21**, No. 4, pp.606-615, May 2003.
41. L. Mhamdi and M. Hamdi, Output queued switch emulation by a one-cell-internally buffered crossbar switch, *IEEE GLOBECOM*, **7**, pp. 3688-3693, Dec. 2003.
42. F. M. Chiussi and A. Francini, A distributed scheduling architecture for scalable packet switches, *IEEE J. Select. Areas Commun.*, pp. 2665-2683, Dec. 2000.

7

Dual Scheduling Algorithm in a Generalized Switch: Asymptotic Optimality and Throughput Optimality

Lijun Chen, Steven H. Low, and John C. Doyle

Engineering & Applied Science Division, California Institute of Technology
Pasadena, CA 91125, USA chen@cds., slow@, doyle@cds.caltech.edu

The generalized switch is a model of a queueing system in which parallel servers are interdependent and have time-varying service capabilities. It includes as special cases the model of data scheduling in a wireless network and the input-queued crossbar switch model. A finite set S of queues (users) are served in discrete time by a switch. Switch state h follows a discrete-time, finite-state Markov chain. At each time slot t, the switch can choose a scheduling decision m from a finite set M, which captures the interdependency among the servers specifying which subsets of servers can be active simultaneously. Each scheduling decision has the associated vector of service rates $\tilde{r}^m(h(t))$ at which queues are served, where $h(t)$ denotes the switch state at time t.

This article considers the dual scheduling algorithm that uses rate control and queue-length based scheduling to allocate resources for a generalized switch. We first consider a saturated system in which each user has an infinite amount of data to be served. We prove the asymptotic optimality of the dual scheduling algorithm for such a system, which says that the vector of average service rates of the scheduling algorithm maximizes some aggregate concave utility functions. As the fairness objectives can be achieved by appropriately choosing utility functions, the asymptotic optimality establishes the fairness properties of the dual scheduling algorithm. We next consider a system with exogenous arrivals, *i.e.* data flows of finite size arrive at the system randomly. For such a system, we propose a modified dual scheduling algorithm that stabilizes the system whenever the input rates are within the feasible rate region and is then throughput-optimal, *i.e.* achieves 100% throughput.

The dual scheduling algorithm motivates a new architecture for scheduling, in which an additional queue is introduced to interface the user data queue and the time-varying server and to modulate the scheduling process, so as to achieve different performance objectives. Further research would include scheduling with Quality of Service guarantees with the dual scheduler, and its application and implementation in various versions of the generalized switch model.

7.1 Introduction

We consider a general model where a set S of queues (users) are served in discrete time by a generalized switch, as defined in [1]. The generalized switch can be viewed as a discrete-time, interdependent parallel server system. The servers are interdependent in that they cannot provide service simultaneously, and the dependency among them is reflected on the constraints that specify which subsets of servers can be active at the same time. Switch state h follows a discrete-time, finite-state Markov chain. At each time slot t, the switch can choose a scheduling decision m from a finite set M, which captures the constraints imposed by the interdependency among the servers. Each scheduling decision has the associated vector of service rates $\tilde{r}^m(h(t))$ at which queues are served, where $h(t)$ is the switch state at time t.

The generalized switch model has many applications in communication networks. For example, in cellular networks in the downlink, servers correspond to the wireless links from the base stations to the users, and the constraint is that each base station can transmit to at most one of the users and each user can be served by at most one of the base stations in each time slot [2, 3]. Other examples include multi-hop wireless networks where each wireless link can be viewed as a server and the constraints disallow simultaneous transmission of neighboring links due to interference [4–6]. It also includes as a single-state special case input-queued crossbar switch where a server corresponds to each input–output port pair, and the constraint is that each input port transmits to exactly one of the output ports and each output port receives from exactly one of the input ports at any time [7]. The same model can extend to handle the packet switch in wireless network, where the switch state (*i.e.* wireless line rates) is supposedly time-varying.

For such a generalized switch system with time-varying state, the service rate that can be offered to the users (queues) is both user-dependent and time-dependent. This, on one hand, opens up the possibility to use state-aware scheduling strategies, *i.e.* to exploit service variations to increase the throughput. On the other hand, the parallel servers are interdependent, and to serve (schedule) always the user with the highest potential rate maximizes overall throughput but usually results in the starvation of some users. So, we need to trade off throughput for fairness. However, the time-varying nature of the generalized switch, coupled with the user-dependent service rate and unknown data arrival, makes it very challenging to design scheduling policies to fulfil fairness and throughput requirements, as well as other performance objectives.

There exists lots of work on scheduling with different performance objectives for different versions of the generalized switch model. For fair scheduling, in the context of cellular network in the downlink, one of the principal policies is the Proportional Fair Scheduler of Qualcomm High Data Rate system [2,3], which schedules the user with the largest ratio of the current achievable rate to the exponentially smoothed throughput. This scheduling algorithm has been shown to maximize the sum of the logarithm utilities of the long-run average data rates provided to the users [8–10], and thus achieve proportional fairness [11]. The generalization of the proportional fair scheduling algorithm to any concave utility function for a generalized switch has been studied[1] [12]. Other work on fair scheduling includes [13–15]. For throughput-optimal scheduling that attains the maximum stability region of the system, one of the principal policies is MaxWeight scheduling in the context of wireless networks [4–6, 16–19], and in the context of input-queued switches [7]. The stability region of a scheduling policy is the set of mean flow rate vectors such that the queue-length process is stable under this policy. The throughput-optimal scheduling has its origin in [4–6], where it is shown that allocating resources to maximize a queue-length-weighted sum of rates is a stabilizing policy under any sustainable flows. However, there is no fairness guarantee with throughput-optimal scheduling.

In this article, we study the dual scheduling algorithms for the generalized switch (see [20] for preliminary results in the context of cellular networks in the downlink). These algorithms are motivated by the dual subgradient algorithm of convex optimization problems [21,22]. With an additional queue (termed M-queue) being intro duced for each user, the dual scheduling algorithm is a combination of rate control (of the M-queue) and M-queue-length-based scheduling. The rate control algorithm is motivated by utility framework for TCP congestion control [23,24], which shows that various TCP congestion control protocols can be interpreted as distributed primal-dual algorithms to solve aggregate network utility maximization. The queue-length-based scheduling takes the form of a simple throughput-optimal scheduling. As such, while the queue-length-based scheduling part keeps maximizing the throughput, the rate-control part modulates the scheduling process by choosing appropriate utility functions, so as to achieve various performance objectives.

Section 7.2 presents details of the system model and the dual scheduling algorithm for the generalized switch. In Section 7.3, we consider a saturated system in which each user has an infinite amount of data to be served. For such a system, fairness among the users is presumably the most important concern. We present a dual scheduling algorithm, which extends the algorithm studied in [25], and prove its asymptotic optimality, which says that the vector of average service rates of the scheduling algorithm maximizes some aggregate concave utilities of the users. As is well known, the fairness objectives can be achieved by choosing the utility functions appropriately. So, the asymptotic optimality establishes the fairness properties of the dual scheduling algorithm.

We next consider in Section 7.4 the system with exogenous arrivals, *i.e.* data flows of finite size arrive at the system randomly. For such a system, one of the performance properties of particular concern for any scheduling algorithm is its stability rate region within which the flows can be stably supported and the response time remains finite. We consider the traffic at flow level, but include packet-level dynamics, *i.e.* the precise operation of the scheduling mechanism. We propose another dual scheduling algorithm and show that it stabilizes the system whenever the arrival rates are within the feasible rate region and is then throughput-optimal. Compared with other throughput-optimal scheduling policies, the dual throughput-optimal scheduling algorithm provides some weighted fairness among the users at flow level.

---

[1] We call this type of scheduling policy a primal scheduling algorithm, since it can be seen as a gradient algorithm to solve the concave utility maximization problem directly.

The dual scheduling algorithm motivates a new architecture for scheduling in the generalized switch, in which an additional queue is introduced to interface the user data queue and the time-varying server and to modulate the scheduling process, so as to achieve different performance objectives such as fairness, and maximum throughput, etc. In Section 7.5, we will briefly discuss some implementation issues and advantages of the dual scheduler, and Quality of Service scheduling in generalized switches.

7.2 System Model

We consider a queueing system where a finite set S of parallel queues (users), indexed by s, are served by a generalized switch. The generalized switch can be abstracted as an interdependent parallel server system. The servers are interdependent in that they cannot provide service simultaneously, and the dependency among them is reflected on the constraints that specify which subsets of servers can be active at the same time. For convenience, we use a "dependency" graph G to capture this interdependency. Each vertex in G represents a server, and an edge between two vertices means the corresponding servers cannot be active simultaneously. Thus, only those servers in an independent set[2] of the dependency graph can be active at the same time. We denote the set of independent sets by M, with each element indexed by m.

The system operates in discrete time $t = 0, 1, 2, \cdots$. By convention, we choose the duration of a time slot as the *unit* of time, and identify time t with the unit time interval $[t, t + 1)$. The switch has a finite set H of states. The switch state is fixed in one of the states $h \in H$ within a time slot but varies across slots according to an irreducible finite-state Markov chain. Corresponding to the switch state h, the service rate to user s is $r_s(h)$ packets per time slot when the switch servers only s, and the service rate vectors $\tilde{r}^m(h),\ m \in M$ that can be offered to the users are

$$\tilde{r}_s^m(h) = \begin{cases} r_s(h) & \text{if } s \in m \\ 0 & \text{otherwise} \end{cases}$$

By standard time-sharing argument, the feasible rate region $\Pi(h)$ in switch state h is defined to be the convex hull of these rate vectors [26]

$$\Pi(h) := \left\{ \tilde{r} : \tilde{r} = \sum_{i=1}^{M} t_i \tilde{r}^i(h),\ t_i \geq 0,\ \sum_{i=1}^{M} t_i = 1 \right\} \tag{7.1}$$

where we slightly abuse the notation and let M also denote the size of the set M. Let the switch state distribution be $d(h)$, we further define the mean feasible rate (capacity) region as

[2] An independent set of vertices is defined as a set of vertices that have no edges between each other. An empty set is an interdependent set.

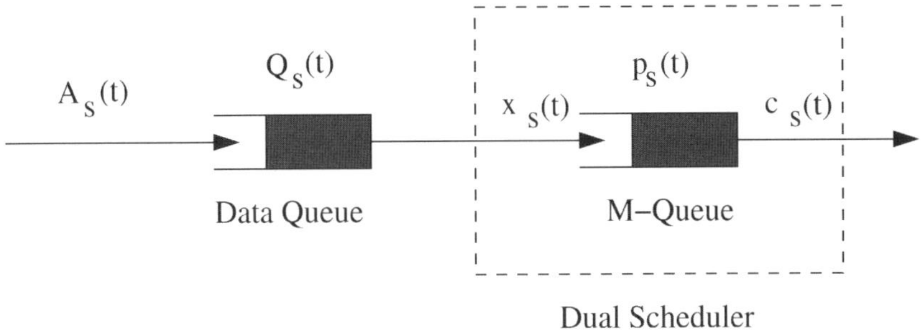

Figure 7.1. The architecture of the dual scheduler

$$\overline{\Pi} = \left\{ \overline{r} : \overline{r} = \sum_{h \in H} d(h)\tilde{r}(h), \tilde{r}(h) \in \Pi(h) \right\} \tag{7.2}$$

This mean rate region is a closed convex set, and is the best feasible rate region the system can support on average.

7.2.1 Queue Length Dynamics

Figure 7.1 shows the architecture of the dual scheduler from the perspective of one user. The system keeps separate *data queues* for the users to buffer the data intended to them. In addition, another queue, called M-queue, is introduced for each user. The M-queue interfaces the data queue and the time-varying server, in that the data will depart from the data queue to enter the M-queue, and the server will directly serve the M-queue.

Denote the size of the data queue and M-queue for user s at the beginning of the time slot t by $Q_s(t)$ and $p_s(t)$ respectively, the number of arrivals to the data queue and M-queue of user s in time slot t by $A_s(t)$ and $x_s(t)$ respectively, and the amount of service offered to the M-queue of user s in time slot t by $c_s(t)$. The evolutions of the data queue and M-queue length for user s are given by

$$Q_s(t+1) = Q_s(t) + A_s(t) - x_s(t) \tag{7.3}$$

$$p_s(t+1) = [p_s(t) + x_s(t) - c_s(t)]^+ \tag{7.4}$$

where '+' denotes the projection onto the set $\Re^+$ of non-negative real numbers.

We further introduce a small parameter $\gamma > 0$, and for convenience, define a new quantity $q_s(t) = \gamma p_s(t)$ for each user s. In Section 7.3 we will see that γ characterizes the asymptotic optimality and fairness of the dual scheduling algorithm. We call q the scaled queue-length, since it is the M-queue length scaled by γ. By Equation (7.4), the evolution of the scaled queue-length is given by

$$q_s(t+1) = [q_s(t) + \gamma(x_s(t) - c_s(t))]^+ \tag{7.5}$$

With the dual scheduling algorithm, the system controls the arrival rate into the M-queues and determines service rates offered to the M-queues based on queue-length.

7.2.2 Dual Scheduling Algorithm

We assume that each user s attains a utility $U_s(x_s)$ when its arrival rate at the M-queue is x_s packets per time slot. $U_s(\cdot)$ may be dependent of data queue size Q_s, but is assumed to be continuously differentiable, increasing and strictly concave with respect to x_s. In time slot t, given the current M-queue length $p_s(t)$, the maximal arrival rate to the M-queue of user s is specified as follows:

$$x_s(t) = \min\left\{U_s'^{-1}(\gamma p_s(t)), \alpha_s\right\} = \min\left\{U_s'^{-1}(q_s(t)), \alpha_s\right\} \tag{7.6}$$

where $\alpha_s > \max_h r_s(h)$ is the upper bound specified on the arrival rate, and thus $x_s(t)$ maximizes $U_s(x_s) - q_s x_s$ over $0 \leq x \leq \alpha_s$. Note that we choose a packet of equal length to the unit of data. x_s will be rounded to the closest integer automatically.

We now consider service allocation. In time slot t, given the current M-queue length $p(t)$, the switch selects a (physical) service rate vector[3]

$$c(t) \in \arg \max_{c \in \Pi(h(t))} p(t)^T c = \arg \max_{c \in \Pi(h(t))} q(t)^T c \tag{7.7}$$

where we will always pick an extreme point maximizer[4]. Equation (7.7) takes the form of simple throughput-optimal scheduling as proposed in [4, 5], which schedules the transmissions dynamically based only on current system backlog and switch state.

Equations (7.3)–(7.7) define the dual scheduling algorithm. When the M-queue length process is stable, x_s *will be the service rate offered to user* s. This scheduling algorithm can be seen as motivated by the dual subgradient algorithm of the concave maximization problem $\max_x \sum_s U_s(x_s)$, and is a combination of rate control [23, 24] and queue-length-based scheduling. As the queue-length-based scheduling part keeps maximizing the throughput, the rate-control part modulates the scheduling process by choosing appropriate utility functions, so as to achieve various performance objectives.

Given a scheduling algorithm, two important issues that need to be addressed are to characterize its fairness property and its stability region. The fairness property governs resource allocation among the competing users, and the stability region determines the efficiency of the scheduling algorithm as a whole. We will study them in the next two sections, respectively.

[3] We call the service rate allocated to the M-queue the physical service rate, in order to distinguish from the service rate received by the user data queue which will be x_s if M-queue is table.

[4] A point in a convex set is an extreme point if it cannot be written as a convex combination of other points in the convex set.

7.3 Asymptotic Optimality and Fairness

In this section, we consider a saturated system in which each user has an infinite amount of data to be served, *i.e.* the user data queue is infinitely backlogged. So, the data queue is irrelevant and the choice of utility function $U_s(\cdot)$ is independent of Q_s. We will show that the dual scheduling algorithm maximizes some aggregate concave utilities and establish its fairness properties through its asymptotic optimality.

7.3.1 An Ideal Reference System

Before proceeding, let us first define an ideal reference system problem:

$$\max_{x_s \geq 0, c_s \geq 0} \sum_s U_s(x_s) \tag{7.8}$$

$$\text{subject to } x \leq c \;\&\; c \in \overline{\Pi} \tag{7.9}$$

The first constraint says that the arrival rate at the M-queues should not exceed the physical service rate. The second constraint says that the physical service rate should be in the mean rate region, which is the best feasible rate region the system can support. We will characterize the performance of the dual scheduling algorithm with respect to this reference system.

Proposition 1. *The solution x^* to problem (7.8),(7.9) exists and is unique.*

Proof. The proof is trivial, since the objective function is strictly concave and the constraint set is a closed, convex set [21].

Consider the dual problem of the reference system problem (7.8)-(7.9)

$$\min_{u \geq 0} D(u) \tag{7.10}$$

with partial dual function

$$D(u) = \max_{x_s \geq 0, c_s \geq 0} \sum_s U_s(x_s) - u^T(x - c) \tag{7.11}$$

$$\text{subject to } c \in \overline{\Pi} \tag{7.12}$$

where we relax only the constraint $x \leq c$ by introducing Lagrange multiplier u.

Proposition 2. *The solution u^* to dual problem (7.10) exists. Moreover, there is no dual gap between the primal problem (7.8),(7.9) and the dual problem (7.10).*

Proof. The proof is trivial, since problem (7.8),(7.9) is a convex optimization problem [21].

Having established the properties of the ideal reference system problem and its dual, in the next subsection we will characterize the dual scheduling algorithm with respect to them.

Remark 1: Roughly speaking, the primal scheduling algorithm is a scheduling policy whose vector of average service rates solves the problem

$$\max_{x_s \geq 0} \sum_s U_s(x_s)$$
$$\text{subject to } \; x \in \overline{\Pi}$$

This problem is equivalent to problem (7.8),(7.9), since mathematically c can be seen as an auxiliary variable. The primal scheduling algorithm can be seen as being motivated by the gradient algorithm to solve this problem [12], while the dual scheduling algorithm can be seen as being motivated by the dual gradient algorithm to solve the same problem.

7.3.2 Stochastic Stability

Note that M-queue length $p(t)$ (and scaled queue-length $q(t)$) evolves according to a discrete-time, discrete-space Markov chain. We first show that this Markov chain is stable, *i.e.* the queue-length process reaches a steady state and does not go unbounded to infinity. It is easy to check that the Markov chain has a countable state space, but is not necessarily irreducible. In such a general case, the state space is partitioned in transient set T and different recurrent classes R_i. We define *the system to be stable if all recurrent states are positive recurrent and the Markov process hits the recurrent states with probability one* [4]. This will guarantee that the Markov chain will be absorbed/reduced into some recurrent class, and the positive recurrence ensures the ergodicity of the Markov chain over this class.

Theorem 1. *The Markov chains described by Equations (7.4) and (7.5) are stable.*

Proof. Consider the the Lyapunov function $V(q) = \|q - u^*\|_2^2$. By Equations (7.5)–(7.7) and defining $g(q) = c(q) - x(q)$, we have

$$\begin{aligned}
E[\Delta V_t(q)|q] &= E[V(q(t+1)) - V(q(t)) \mid q(t) = q] \\
&= E[V([q(t) - \gamma g(q(t))]^+) - V(q(t)) \mid q(t) = q] \\
&\leq E[V(q(t) - \gamma g(q(t))) - V(q(t)) \mid q(t) = q] \\
&= E[-\gamma g(q(t))^T (2(q(t) - u^*) - \gamma g(q(t))) \mid q(t) = q] \\
&= 2\gamma \overline{g}(q)^T (u^* - q) + \gamma^2 E[\|g(q(t))\|_2^2 \mid q(t) = q] \\
&\leq 2\gamma \overline{g}(q)^T (u^* - q) + \gamma^2 G^2
\end{aligned}$$

where G is the upper bound of the norm of $g(q(t))$, and

$$\overline{g}(q) = \overline{c}(q) - x(q) \;\text{ with }\; \overline{c}(q) \in \arg\max_{c \in \overline{\Pi}} q^T c \tag{7.13}$$

It is easy to check that $\overline{g}(q)$ is a subgradient[5] of the dual function $D(q)$ at point q, thus

$$\overline{g}(q)^T(u^* - q) \leq D(u^*) - D(q)$$

So,

$$E[\Delta V_t(q)|q] \leq 2\gamma(D(u^*) - D(q)) + \gamma^2 G^2$$

Note that $D(q)$ is a continuous function. Let

$$\delta = \max_{D(q)-D(u^*)\leq\gamma G^2} \|q - u^*\|_2$$

and define $\mathcal{A} = \{q : \|q - u^*\|_2 \leq \delta\}$. We can get

$$E[\Delta V_t(q)|q] \leq -\gamma^2 G^2 \mathcal{I}_{q\in\mathcal{A}^c} + \gamma^2 G^2 \mathcal{I}_{q\in\mathcal{A}}$$

where $\mathcal{I}$ is the index function. Thus, by *Theorem 3.1* in [4], which is a trivial extension of Foster's criterion for irreducible chain [27], the Markov chain $q(t)$ is stable. Since the M-queue length $p(t) = \gamma q(t)$, the Markov chain $p(t)$ is also stable.

The above proof shows that the distance to the optimal u^* has negative conditional mean drift for all scaled queue-lengths that have sufficiently large distance to u^*, and implies that the scaled queue-length will stay near u^* when γ is small enough.

Remark 2: We can make the Markov chain $p(t)$ irreducible over its state space, by making the arrival $x(t)$ a random variable with mean $\min\{U_s'^{-1}(\gamma p_s(t)), \alpha_s\}$, as assumed in [25]. We can also make the system reach a specific state infinitely often with finite mean recurrence times, which will ensure that the system reduces to one recurrent class whatever the initial state is.

7.3.3 Asymptotic Optimality and Fairness

In this subsection, we will prove the asymptotic optimality of the dual scheduling algorithm in terms of dual and primal functions of the reference system problem (7.8),(7.9).

Theorem 2. *The dual scheduling algorithm (7.4)–(7.7) converges statistically to within a small neighborhood of the optimal value $D(u^*)$, i.e.*

$$D(u^*) \leq D(E[q(\infty)]) \leq D(u^*) + \gamma G^2/2 \qquad (7.14)$$

where $q(\infty)$ is notation used to denote the state of the Markov chain $q(t)$ in the steady state.

---

[5] Given a convex function $f : \mathcal{R}^n \mapsto \mathcal{R}$, a vector $d \in \mathcal{R}^n$ is a subgradient of f at a point $u \in \mathcal{R}^n$ if $f(v) \geq f(u) + (v-u)^T d,\ v \in \mathcal{R}^n$ [21,22].

Proof. The first inequality $D(u^*) \leq D(p)$ always holds, since $D(u^*)$ is the minimum of the dual function $D(u)$.

Now we prove the second inequality. From the proof of Theorem 1, we have

$$\begin{aligned} E[\Delta V_t(q)|q] &= E[V(q(t+1)) - V(q(t)) \mid q(t) = q] \\ &\leq 2\gamma(D(u^*) - D(q)) + \gamma^2 G^2 \end{aligned}$$

Taking expectation over q, we get

$$\begin{aligned} E[\Delta V_t(q)] &= E[V(q(t+1)) - V(q(t))] \\ &\leq 2\gamma(D(u^*) - E[D(q)]) + \gamma^2 G^2 \end{aligned}$$

Taking summation from $\tau = 0$ to $\tau = t - 1$, we obtain

$$E[V(q(t))] \leq E[V(q(0))] - 2\gamma \sum_{\tau=0}^{t-1} E[D(q(\tau))] + 2\gamma t D(u^*) + t\gamma^2 G^2$$

Since $E[V(q(t))] \geq 0$, we have

$$2\gamma \sum_{\tau=0}^{t-1} E[D(q(\tau))] - 2\gamma t D(u^*) \leq E[V(q(0))] + t\gamma^2 G^2$$

>From this inequality we obtain

$$\frac{1}{t} \sum_{\tau=0}^{t-1} E[D(q(\tau))] - D(u^*) \leq \frac{E[V(q(0))] + t\gamma^2 G^2}{2t\gamma}$$

Note that $q(t)$ is stationary and ergodic in some steady state by Theorem 1, and so is $D(q(t))$. Thus,

$$\lim_{t \to \infty} \frac{1}{t} \sum_{\tau=0}^{t-1} E[D(q(\tau))] = E[D(q(\infty))]$$

So,

$$E[D(q(\infty))] - D(u^*) \leq \gamma G^2/2$$

Since $D(q)$ is a convex function, by Jensen's inequality,

$$D(E[q(\infty)]) - D(u^*) \leq \gamma G^2/2$$

i.e. the algorithm converges statistically to within $\gamma G^2/2$ of the optimal value $D(u^*)$.

Since $D(q)$ is a continuous function, Theorem 2 implies that the scaled queue-length q approaches u^* statistically when γ is small enough.

Corollary 1. *$x(t)$ is a stable Markov chain. Moreover, the average arrival rates $E[x(\infty)] \in \overline{\Pi}$, where $x(\infty)$ denotes the state of the process $x(t)$ in the steady state.*

Proof. $x(t)$ is a deterministic, finite-value function of $q(t)$. $x(t)$ is a stable Markov chain, since $q(t)$ is. $E[x(\infty)] \in \overline{\Pi}$, otherwise the average scaled queue-length $E[q(\infty)]$ will go unbounded, which contradicts Theorem 2.

Theorem 3. *Let $P(x)$ be the primal function of the reference system problem (7.8),(7.9). The dual scheduling algorithm (7.4)–(7.7) converges statistically to within a small neighborhood of the optimal value $P(x^*)$, i.e.*

$$P(x^*) \geq P(E[x(\infty)]) \geq P(x^*) - \frac{\gamma G^2}{2} \tag{7.15}$$

Proof. The first inequality $P(x^*) \geq P(E[x(\infty)])$ holds, since $E[x(\infty)] \in \overline{\Pi}$.

Now we prove the second inequality. By Equation (7.5), we have

$$\begin{aligned}
& E[||q(t+1)||_2^2 | q(t)] \\
&= E[||[q(t) - \gamma g(q(t))]^+||_2^2 | q(t)] \\
&\leq E[||q(t) - \gamma g(q(t))||_2^2 | q(t)] \\
&= ||q(t)||_2^2 - 2\gamma \overline{g}(q(t))^T q(t) + \gamma^2 E[||g(q(t))||_2^2 | q(t)] \\
&= ||q(t)||_2^2 + 2\gamma \sum_s U_s(x_s(t)) - 2\gamma \sum_s (U_s(x_s(t)) - q_s(t) x_s(t)) \\
&\quad -2\gamma \sum_s q_s(t) \overline{c}_s(t) + \gamma^2 E[||g(p(t))||_2^2 | q(t)] \\
&\leq ||q(t)||_2^2 + 2\gamma \sum_s U_s(x_s(t)) - 2\gamma \sum_s (U_s(x_s^*) - q_s(t) x_s^*) \\
&\quad -2\gamma \sum_s q_s(t) \overline{c}_s(t) + \gamma^2 E[||g(p(t))||_2^2 | q(t)] \\
&= ||q(t)||_2^2 + 2\gamma P(x(t)) - 2\gamma P(x_s^*) \\
&\quad -2\gamma \sum_s q_s(t)(\overline{c}_s(t) - x_s^*) + \gamma^2 E[||g(q(t))||_2^2 | q(t)] \\
&\leq ||q(t)||_2^2 + 2\gamma P(x(t)) - 2\gamma P(x^*) + \gamma^2 E[||g(q(t))||_2^2 | q(t)] \\
&\leq ||q(t)||_2^2 + 2\gamma P(x(t)) - 2\gamma P(x^*) + \gamma^2 G^2
\end{aligned}$$

where $\overline{g}(q)$ is defined as in Equation (7.13). The second inequality follows from the fact that $x_s(t)$ is the maximizer of $\max_{x_s}(U_s(x_s) - q_s x_s)$, and the third inequality follows from the fact that $\overline{c}(t)$ is the maximizer in Equation (7.13) and $x^* \in \overline{\Pi}$.

Taking expectation over q, we get

$$E[||q(t+1)||_2^2] \leq E[||q(t)||_2^2] + 2\gamma E[P(x(t))] - 2\gamma P(x^*) + \gamma^2 G^2$$

Applying the inequalities recursively, we obtain

$$E[||p(t)||_2^2] \leq E[||p(0)||_2^2] + 2\gamma \sum_{\tau=0}^{t-1} (E[P(x(\tau))] - P(x^*)) + t\gamma^2 G^2$$

Since $E[||p(t)||_2^2] \geq 0$, we have

$$2\gamma \sum_{\tau=0}^{t-1} (E[P(x(\tau))] - P(x^*)) \geq -E[||p(0)||_2^2] - t\gamma^2 G^2$$

>From this inequality we obtain

$$\frac{1}{t} \sum_{\tau=0}^{t-1} E[P(x(\tau))] - P(x^*) \geq \frac{-E[||p(0)||_2^2] - t\gamma^2 G^2}{2t\gamma}$$

Note that $x(t)$ is stationary and ergodic in some steady state by Corollary 1. Thus,

$$\lim_{t \to \infty} \frac{1}{t} \sum_{\tau=0}^{t-1} E[P(x(\tau))] = E[P(x(\infty))]$$

So,

$$E[(P(x(\infty))] - P(x^*) \geq -\frac{\gamma G^2}{2}$$

Since P is a concave function, by Jenson's inequality,

$$P(E[x(\infty)]) - P(x^*) \geq -\frac{\gamma G^2}{2}$$

i.e. the algorithm converges statistically to within $\gamma G^2/2$ of the optimal value $P(x^*)$.

Since $P(x)$ is a continuous function, Theorems 3 implies that the average arrival rates at the M-queues approaches the optimal of the ideal reference system (7.8),(7.9) when γ is small enough. Note that, when the M-queue length is stable, x will be the service rates offered to the users. Theorems 2 and 3 show that, surprisingly, the vector of average service rates offered by the dual scheduling algorithm (7.4)–(7.7) approximately solves the ideal reference system problem, which is to maximize the aggregate concave utilities over the best feasible rate region that the network can support.

As is well known, the fairness objectives can be achieved by choosing the concave objective functions appropriately. So, the asymptotic optimality establishes the fairness properties of the dual scheduling algorithm. For example, if we choose logarithm utility function $U_s(x_s) = \log(x_s)$, the dual scheduler will achieve proportional fairness [11,25].

7.4 Throughput-optimal Scheduling

The results presented in the previous section assume a saturated system in which each user has an infinite amount of data to be served. In reality, however, data flows of finite size arrive at the system for each user according to some arrival process. For such a system, one of the properties of particular concern for a scheduling algorithm is its stability rate region within which the flows can be stably supported and the response time remains finite, since it characterizes the throughput performance of the scheduling algorithm as a whole. In this section, we will consider a scheduling policy which maximizes the system throughput while providing some weighted fairness among users at flow level.

We will model the traffic in the system at the level of flows. We assume that the arrival process of data flows for each user s is a stationary point process with intensity ν_s, and the size of the flows to be served is drawn independently from a general distribution with mean $1/\mu_s$ and finite second moment. We denote by $\rho_s = \nu_s/\mu_s$ the traffic intensity offered to user s. The data queue length (unfinished work) $Q(t)$ is a stochastic process which increases as new flows come and decreases as date is transferred into the M-queues. We are concerned with the stability of $Q(t)$, and derive conditions on the traffic intensity ρ for which, starting from any initial state, the amount of data (or number of flows) to be served remains finite with probability 1. Obviously, the following condition is necessary

$$\rho \in \text{int}(\overline{\Pi}) \tag{7.16}$$

where $\text{int}(\overline{\Pi})$ denotes the interior of convex set $\overline{\Pi}$. This condition for stability is not sufficient for general scheduling policies. We call a scheduling algorithm *throughput-optimal* if condition (7.16) is also sufficient under this algorithm.

7.4.1 Throughput Optimality and Fairness

We would like the dual scheduling algorithm discussed in the previous section to be a throughput-optimal policy. However, fairness and maximum throughput are not directly compatible performance objectives. On one hand, as we can see in [4–6], there is no guarantee of fairness with throughput-optimal scheduling. On the other hand, fair scheduling may not be stable for some flows within the feasible rate region; see the discussion in [28] for proportional fair scheduling algorithm in cellular network in the downlink. The main reason why fair scheduling may not be a stabilizing policy is because it is usually traffic-independent. For example, in the proportional fair scheduling algorithm, the scheduling decision is made based on the current achievable rate and the mean throughput but not on the real traffic intended by the users.

Given the above consideration, a throughput-optimal scheduling policy should be traffic-dependent. This motivates us to consider a scheduling algorithm whose vector of service rates x solves the following utility maximization:

$$\max_{x_s \geq 0, c_s \geq 0} \sum_s Q_s U_s(x_s) \tag{7.17}$$

$$\text{subject to } x \leq c \,\&\, c \in \overline{\Pi} \tag{7.18}$$

where $U_s(\cdot)$ is again independent of the data queue length Q_s. However, we cannot solve problem (7.17),(7.18) directly, since the system only knows the current switch state but not its statistics.

By the assumptions about data flows for each user, the arrival at the data queues $A_s(t)$ is i.i.d. across time slots and with mean $\rho_s = E[A_s(t)]$ and finite second moment. The evolution of unfinished work $Q(t)$ and the modified dual scheduling algorithm are given by

$$Q_s(t+1) = [Q_s(t) + A_s(t) - x_s(t)]^+ \tag{7.19}$$
$$q_s(t+1) = [q_s(t) + \gamma(\tilde{x}_s(t) - c_s(t))]^+ \tag{7.20}$$
$$x_s(t) = \min\{U_s'^{-1}(q_s(t)/Q_s(t)), \alpha_s\} \tag{7.21}$$
$$c(t) \in \arg \max_{c \in \Pi(h(t))} q(t)^T c \tag{7.22}$$

where $\tilde{x}_s(t) = \min\{Q_s(t) + A_s(t), x_s(t)\}$. Equations (7.20)–(7.22) are the modified dual scheduling algorithm motivated by the utility maximization (7.17),(7.18). It makes the scheduling decision based only on current backlog and switch state. Note that if the unfinished work $Q(t)$ freezes, this algorithm will solve the utility maximization approximately.

7.4.2 Optimality Proof

It is easy to verify that the pair $(Q(t), q(t))$[6] evolves according to a discrete-time, discrete-space Markov chain. The following theorem proves the stability of this Markov chain under any sustainable flows, and shows that the modified dual scheduling algorithm is a throughput-optimal scheduling policy.

Theorem 4. *The Markov chain $Y(t) = (Q(t), q(t))$ is stable iff traffic condition (7.16) is satisfied.*

Before we prove Theorem 4, we first introduce two lemmas. Consider Lyapunov function

$$V(Y) = V_1(Y) + V_2(Y)$$

where $V_1(Y) = \sum_s b_s Q_s^2 = \sum_s U_s'((1+\epsilon)\rho_s)Q_s^2$, and $V_2(Y) = \sum_s \frac{q_s^2}{\gamma}$. ϵ is a small positive constant that will be decided later on. Note that $b_s = U_s'((1+\epsilon)\rho_s) > 0$, since $U_s(\cdot)$ is an increasing function. Also, note that $U_s'(\cdot)$ is a decreasing function.

Lemma 1. *For any $\rho \in int(\overline{\Pi})$, there exists a $\epsilon > 0$ such that the following inequality holds:*

[6] Throughout this section, for convenience, we will use scaled queue-length $q(t)$ rather than the M-queue length $p(t)$ to establish the stability and optimality properties of the dual scheduling algorithm. This makes no difference, since $q(t)$ is $p(t)$ scaled by a constant and all the properties of $q(t)$ apply to $p(t)$ and vice versa.

$$E[V_1(Y(t+1)) - V_1(Y(t))|Y(t)]$$
$$\leq -2\epsilon \sum_s b_s \rho_s Q_s(t) + 2 \sum_s q_s(t)((1+\epsilon)\rho_s - x_s(t)) + T_1$$

where T_1 is a positive constant.

Proof. For any $\rho \in \text{int}(\overline{\Pi})$, there exists a small number $\epsilon > 0$ such that $(1+\epsilon)\rho \in \text{int}(\overline{\Pi})$, since $\overline{\Pi}$ is a closed convex set and $0 \in \overline{\Pi}$. By equation (7.19), we have

$$\begin{aligned}
& E[V_1(Y(t+1)) - V_1(Y(t))|Y(t)] \\
&= \sum_s b_s E[([Q_s(t) + A_s(t) - x_s(t)]^+)^2 - Q_s^2(t)|Y(t)] \\
&\leq \sum_s b_s E[(Q_s(t) + A_s(t) - x_s(t))^2 - Q_s^2(t)|Y(t)] \\
&= \sum_s b_s E[2Q_s(t)(A_s(t) - x_s(t))|Y(t)] + \sum_s b_s E[(A_s(t) - x_s(t))^2|Y(t)] \\
&\leq \sum_s 2b_s Q_s(t)(\rho_s - x_s(t)) + T_1 \\
&= -\sum_s 2\epsilon b_s \rho_s Q_s(t) + \sum_s 2b_s Q_s(t)((1+\epsilon)\rho_s - x_s(t)) + T_1
\end{aligned} \tag{7.23}$$

Here the positive constant T_1 is the upper bound of $\sum_s b_s E[(A_s(t) - x_s(t))^2|Y(t)]$. This can be achieved since $x_s(t)$ is upper bounded by α_s and $E[A_s^2(t)]$ is also bounded.

Now we consider the second term in (7.23). If $(1+\epsilon)\rho_s \geq x_s(t)$,

$$\begin{aligned}
& 2b_s Q_s(t)((1+\epsilon)\rho_s - x_s(t)) \\
&\leq 2Q_s(t)U_s'(x_s(t))((1+\epsilon)\rho_s - x_s(t)) \\
&= 2q_s(t)((1+\epsilon)\rho_s - x_s(t))
\end{aligned}$$

where the first inequality follows from the fact that $U_s'(\cdot)$ is a decreasing function, and the last equality follows from the fact $x_s(t) \leq (1+\epsilon)\rho_s < \alpha_s$ and Equation (7.21). If $(1+\epsilon)\rho_s < x_s(t)$,

$$\begin{aligned}
& 2b_s Q_s(t)((1+\epsilon)\rho_s - x_s(t)) \\
&= 2q_s(t)((1+\epsilon)\rho_s - x_s(t)) + 2(b_s Q_s(t) - q_s(t))((1+\epsilon)\rho_s - x_s(t)) \\
&\leq 2q_s(t)((1+\epsilon)\rho_s - x_s(t)) + 2(Q_s(t)b_s - Q_s(t)U_s'(x(t)))((1+\epsilon)\rho_s - x_s(t)) \\
&\leq 2q_s(t)((1+\epsilon)\rho_s - x_s(t))
\end{aligned}$$

where the first inequality follows from Equation (7.21), and the last inequality follows from the fact that $U_s'(\cdot)$ is a decreasing function.

So, in any case, we get

$$E[V_1(Y(t+1)) - V_1(Y(t))|Y(t)]$$
$$\leq -2\epsilon \sum_s b_s \rho_s Q_s(t) + 2 \sum_s q_s(t)((1+\epsilon)\rho_s - x_s(t)) + T_1$$

Lemma 2. *The following inequality holds:*

$$E[V_2(Y(t+1)) - V_2(Y(t))|Y(t)] \le 2\sum_s q_s(t)(x_s(t) - \overline{c}_s(t)) + T_2$$

where T_2 *is a positive constant, and* $\overline{c}(t) = arg\max_{c\in\overline{\Pi}} q(t)^T c$.

Proof. By Equation (7.20), we have

$$\begin{aligned}
&E[V_2(Y(t+1)) - V_2(Y(t))|Y(t)] \\
&= \sum_s \frac{1}{\gamma} E[([q_s(t) + \gamma(\tilde{x}_s(t) - c_s(t))]^+)^2 - q_s^2(t)|Y(t)] \\
&\le \sum_s \frac{1}{\gamma} E[(q_s(t) + \gamma(x_s(t) - c_s(t)))^2 - q_s^2(t)|Y(t)] \\
&= 2\sum_s E[q_s(t)(x_s(t) - c_s(t))|Y(t)] + \sum_s \gamma E[(x_s(t) - c_s(t))^2|Y(t)] \\
&= 2\sum_s q_s(t)(x_s(t) - \overline{c}_s(t)) + \sum_s \gamma E[(x_s(t) - c_s(t))^2|Y(t)] \\
&= 2\sum_s q_s(t)(x_s(t) - \overline{c}_s(t)) + T_2
\end{aligned}$$

Here the positive constant T_2 is the upper bound of $\sum_s \gamma E[(x_s(t) - c_s(t))^2|Y(t)]$. This can be achieved since both $x_s(t)$ and $c_s(t)$ are bounded.

With Lemmas 1 and 2, we are ready to prove Theorem 4.

Proof. For any $\rho \in \mathrm{int}(\overline{\Pi})$, there exists a small number ϵ such that $(1+\epsilon)\rho, (1+2\epsilon)\rho \in \mathrm{int}(\overline{\Pi})$. Thus, by Lemmas 1 and 2, for any arrival process $A(t)$ with mean ρ we have

$$\begin{aligned}
&E[V(Y(t+1)) - V(Y(t))|Y(t)] \\
&\le -2\epsilon\sum_s b_s\rho_s Q_s(t) + 2\sum_s q_s(t)((1+\epsilon)\rho_s - x_s(t)) + T_1 \\
&\quad +2\sum_s q_s(t)(x_s(t) - \overline{c}_s(t)) + T_2 \\
&= -2\epsilon\sum_s b_s\rho_s Q_s(t) + 2\sum_s q_s(t)((1+\epsilon)\rho_s - \overline{c}_s(t)) + T \\
&= -2\epsilon\sum_s b_s\rho_s Q_s(t) - 2\epsilon\sum_s \rho_s q_s(t) + T \\
&\quad +2\sum_s q_s(t)((1+2\epsilon)\rho_s - \overline{c}_s(t)) \\
&\le -2\epsilon\sum_s b_s\rho_s Q_s(t) - 2\epsilon\sum_s \rho_s q_s(t) + T
\end{aligned}$$

where $T = T_1 + T_2$, and the last inequality follows from the fact that $\overline{c}(t)$ maximizes $q(t)^T c$ over $c \in \overline{\Pi}$.

We can see that when $||Y(t)||_2$ is large enough, the conditional mean drift of Lyapunov function $V(Y(t))$ is negative. Thus, by Theorem 3.1 in [4], the Markov chain $Y(t)$ is stable.

Theorem 4 shows that the system is stable under any sustainable flow $\rho \in \text{int}(\overline{\Pi})$. So, the modified dual scheduling algorithm is a throughput-optimal policy. Note that we prove throughput optimality under a very general assumption of the flow arrival process, while most similar results assume a Poisson arrival with exponentially distributed file size. Also, the above stability result is independent of the value of parameter γ, while in Section 7.3 the parameter γ characterizes the optimality of the dual scheduling algorithm.

7.4.3 Flows with Exponentially Distributed Size

If the flow lengths are exponentially distributed with parameter $1/\mu_s$, the system can be modeled by a Markov chain $(n(t), q(t))$, where $n_s(t)$ is the number of active data flows for user s by the beginning of time slot t. Similarly, we would like the vector of service rates x provided by the scheduling algorithm to solve the following utility maximization:

$$\max_{x_s \geq 0, c_s \geq 0} \sum_s n_s U_s(x_s) \tag{7.24}$$

$$\text{subject to } x \leq c \;\&\; c \in \overline{\Pi} \tag{7.25}$$

Let $a_s(t)$ denote the number of new flows of user s arriving in time slot t, and $d_s(t)$ the number of flows that complete data transfer in time slot t. Thus, $E[a_s(t)] = \nu_s$ and $E[d_s(t)|n(t), q(t)] = \mu_s x_s(t)$. Motivated by the dual subgradient algorithm for the above utility maximization, the evolution of $n(t)$ and the modified dual scheduling algorithm are given by

$$n_s(t+1) = n_s(t) + a_s(t) - d_s(t) \tag{7.26}$$

$$q_s(t+1) = [q_s(t) + \gamma(x_s(t) - c_s(t))]^+ \tag{7.27}$$

$$x_s(t) = \min\{U_s'^{-1}(q_s(t)/n_s(t)), \alpha_s\} \tag{7.28}$$

$$c(t) \in \arg \max_{c \in \Pi(h(t))} q(t)^T c \tag{7.29}$$

Considering Lyapunov function $V(n, q) = \sum_s b_s n_s^2/\mu_s + \sum_s q_s^2/\gamma$ and following the same procedure as that in the proof of Theorem 4, we can directly obtain the following theorem.

Theorem 5. *The Markov chain $(n(t), q(t))$ is stable iff traffic condition (7.16) is satisfied.*

This theorem shows that the system is stable under any sustainable flow $\rho \in \text{int}(\overline{\Pi})$. So, the modified dual scheduling algorithm (7.27)–(7.29) is a throughput-optimal policy. Compared to other throughput-maximum policy, the modified dual scheduling algorithm has the advantage of providing some short-term fairness. To see this, note that Equations (7.27)–(7.29) are the dual subgradient algorithm. If the number of active flows n_s for each user freezes, the above scheduling algorithm will solve the utility maximization (7.24),(7.25). But n_s is time-varying, and the system cannot achieve asymptotic optimality before the change of n_s. Nonetheless, at the level of flows whose time-scale is large compared to the duration of the time slot, the algorithm will provide some level of fairness, since the system evolves along the gradient direction to the optimal.

Note that throughout this section, we consider the system traffic at flow level, but include packet level dynamics, *i.e.* the precise operation of scheduling mechanism. We may also assume "the separation of time-scale"[7], *i.e.* the modified dual scheduling algorithm (7.27)–(7.29) works at fast time-scale and achieves the resource sharing objective (7.24),(7.25) perfectly at flow level. Under this assumption, $n(t)$ is an irreducible Markov chain, and we can choose Lyapunov function $V(n(t)) = \sum_s b_s n_s^2/\mu_s$ to prove that the algorithm is a throughput-optimal policy.

Theorem 6. *Under the assumption of the separation of time-scale, the Markov chain $n(t)$ is stable iff traffic condition (7.16) is satisfied.*

Proof. Let x denote the maximizer of the problem (7.24),(7.25). For any $\rho \in \text{int}(\overline{\Pi})$, there exists a small number ϵ such that $(1+\epsilon)\rho \in \text{int}(\overline{\Pi})$. So, $\sum_s n_s U_s(x_s) \geq \sum_s n_s U_s((1+\epsilon)\rho_s)$. Note that $U_s(\cdot)$ is a strictly concave function, thus

$$\begin{aligned}&\sum_s n_s b_s (x_s - (1+\epsilon)\rho_s)\\ &\geq \sum_s n_s U_s(x_s) - \sum_s n_s U_s((1+\epsilon)\rho_s) \geq 0 \end{aligned} \tag{7.30}$$

Now consider the conditional drift of Lyapunov Function $V(n)$:

$$\begin{aligned}&E[V(n(t+1)) - V(n(t))|n(t)]\\ &= 2\sum_s \frac{b_s}{\mu_s} E[n_s(t)(a_s(t) - d_s(t))|n(t)] + \sum_s \frac{b_s}{\mu_s} E[(a_s(t) - d_s(t))^2|n(t)]\\ &\leq 2\sum_s b_s n_s(t)(\rho_s - x_s) + T_3\end{aligned}$$

where positive constant T_3 is the upper bound of $\sum_s \frac{b_s}{\mu_s} E[(a_s(t) - d_s(t))^2|n(t)]$. Thus,

[7] Due to its analytical tractability, most researches on user-level performance of scheduling algorithms assume a separation of time-scale [14, 29].

$$
\begin{aligned}
&E[V(n(t+1)) - V(n(t))|n(t)] \\
&\leq -2\epsilon \sum_s b_s\rho_s n_s(t) + T_3 + 2\sum_s b_s n_s(t)((1+\epsilon)\rho_s - x_s) \\
&\leq -2\epsilon \sum_s \rho_s b_s n_s(t) + T_3
\end{aligned}
$$

where the last inequality follows from Equation (7.30). Thus, by Foster's criterion [27], the Markov chain $n(t)$ is stable.

Remark 3: Many throughput-optimal scheduling algorithms for different versions of the generalized switch model have been proposed [1,4–7,16,17,19]. Most of these scheduling policies maximize $\sum_s f_s(Q_s(t))x_s$, where $f_s(\cdot)$ is an explicitly or implicitly defined function of backlog Q_s, in order to achieve different performance objectives or implement different scheduling criteria. However, to our knowledge there does not exist any throughput-optimal policy which maximizes $\sum_s Q_s(t)F_s(x_s)$, where $F_s(\cdot)$ is a function of service rate x_s. It is possible to achieve different performance objectives by different choices of $F_s(\cdot)$. In the modified dual scheduling algorithm, we can say that we introduce M-queue to modulate the scheduling, which implicitly defines a function $F_s(\cdot)$. This modulation will definitely change the dynamics of the scheduling process, and to provide some weighted fairness among the users at flow level is a such consequence. It is interesting to further study the related issues.

7.5 A New Scheduling Architecture

The dual scheduling algorithm motivates a new architecture for scheduling in the generalized switch (please see Figure 7.1 for a pictorial depiction). In this new architecture, a queue, termed M-queue, is introduced to interface the user data queue and the time-varying server. Data will depart from the data queue to enter the M-queue, and the generalized switch serves directly the M-queue. Through controlling the arrival process to the M-queue (or the departure process from the data queue), we can modulate the scheduling process, in order to achieve different performance objectives such as fairness, and maximum throughput, etc.

The dual scheduler would not incur much additional complexity. M-queues are distributed at each user, and can be "virtual" or implemented as physical queues. The control of the M-queue arrival process is also distributed at each user and depends on only the "local" queue length of each user. The dual scheduler provides some advantages over other scheduling algorithms. For example, in the cellular network in the downlink, even though the primary scheduling algorithm can achieve fair resource allocation, it requires to estimate the average throughput of the users, while with the dual algorithm we only need to simply measure the M-queue length. Also, the dynamics of the M-queue ($p(t)$ and $x(t)$) is feedback-controlled, and thus will be relatively smooth, comparing with the dynamics of the switch. So, the dual scheduler can provide a relatively reliable and smooth service to the users, and can behave as a good interface between higher layer protocols and the scheduling at the link layer and ensure a better performance of the higher-layer protocols such as that of TCP congestion control.

To provide Quality of Service in generalized switches is a difficult problem. In the context of wireless networks, the interdependence of wireless links in combination with the time-varying nature of wireless channel makes QoS scheduling fairly challenging and the available results are mostly on stability guarantees. In the context of input-queued crossbar switch, input buffering makes scalable switch design possible but makes QoS guarantees very challenging and again most available results are on maximizing the throughput; see [30] for a review. The dual scheduler might be promising in providing QoS in generalized switches, through carefully designing the M-queue arrival process. Further study is needed on related issues.

7.6 Conclusions

In this article, we consider the dual scheduling algorithm for a generalized switch. For a saturated system, we prove the asymptotic optimality of the dual scheduling algorithm and thus establish its fairness properties. For a system with exogenous arrivals, we propose a modified dual scheduling algorithm, which is throughput-optimal while providing some weighted fairness among the users at the level of flows.

The dual scheduling algorithm motivates a new architecture for scheduling, in which an additional queue is introduced to interface the user data queue and the time-varying server and to modulate the scheduling process, so as to achieve different performance objectives. Further research stemming out of this article includes scheduling with Quality of Service guarantees with the dual scheduler, and its application and implementation in various versions of the generalized switch model.

Acknowledgments

This work is part of the Caltech FAST Project supported by NSF, Caltech Lee Center for Advanced Networking, ARO, AFOSR, DARPA, and Cisco.

References

1. A. L. Stolyar, MaxWeight scheduling in a generalized switch: state space collapse and workload minimization in heavy traffic, *Ann. Appl. Probab.*, **14**(1):1-53, 2004.
2. P. Bender, P. Black, M. Grob, R. Padovani, N. Sindhushayana and A. Viterbi, CDMA/HDR: a bandwidth-efficient high-speed wireless data service for nomadic users, *IEEE Communication Magazine*, **38**(7):70-77, 2000.
3. A. Jalali, R. Padovani and R. Pankaj, Data-throughput of CDMA-HDR a high efficiency data rate personal communication wireless system, *Proc. 50th IEEE VTC*, 2000.
4. L. Tassiulas and A. Ephremides, Stability properties of constrained queueing systems and scheduling policies for maximum throughput in multihop radio networks, *IEEE Transactions on Automatic Control*, **37**(12):1936-1948, December 1992.
5. L. Tassiulas and A. Ephremides, Dynamic server allocation to parallel queues with randomly varying connectivity, *IEEE Transactions on Information Theory*, **39**:466-478, 1993.
6. L. Tassiulas, Scheduling and performance limits of networks with constantly changing topology, *IEEE Transactions on Information Theory*, **43**(3):1067-1073, 1997.
7. N. McKeown, V. Anantharam and J. Walrand, Achieving 100% throughput in an input-queued switch, *Proc. IEEE Infocom*, 1996.
8. H. J. Kushner and P. A. Whiting, Convergence of proportional-fair sharing algorithms under general conditions, *IEEE/ACM Transactions on Wireless Communications*, **3**(4):1250-1259, 2004.
9. V. Subramanian and R. Agrawal, A stochastic approximation analysis of channel condition aware wireless scheduling algorithms, *Proc. INFORMS Telecommun. Conf.*, 2002.
10. D. N. Tse, Multiuser diversity and proportionally fair scheduling, *In preparation*, 2004.
11. F. P. Kelly, Charging and rate control for elastic traffic, *European Transactions on Telecommunications*, **8**:33-37, 1997.
12. A. L. Stolyar, On the asymptotic optimality of the gradient scheduling algorithm for multiuser throughput allocation, *Operation Research*, **53**(1):12-25, 2005.
13. X. Liu, E. Chong and N. Shroff, Opportunistic transmission scheduling with resource-sharing constraints in wireless networks, *IEEE J. Sel. Area Commun.*, **19**(10):2053-2064, 2001.
14. S. Borst, User-level performance of channel-aware scheduling algorithm in wireless data networks, *Proc. IEEE Infocom*, 2003.
15. Y. Liu and E. Knightly, Opportunistic fair scheduling over multiple wireless channels, *Proc. IEEE Infocom*, April 2003.
16. M. Andrews, K. Kumaran, K. Ramanan, A. Stolyar, R. Vijayakumar and P. Whiting, Providing quality of services over shared wireless link, *IEEE Communication Magazine*, Feburary 2001.
17. A. Eryilmaz, R. Srikant and J. Perkins, Stable scheduling policies for fading wireless channels, *IEEE/ACM Transactions on Networking*, 2005.
18. M. Neely, E. Modiano and C. Rohrs, Dynamic power allocation and routing for time varying wireless networks, *Proc. IEEE Infocom*, April 2003.
19. S. Shakkottai and A. Stolyar, Scheduling for multiple flows sharing a time-varying channel: the exponential rules, *Transactions of the AMS*, Series 2, A volume in memory of F. Karpelevich, 2002.
20. L. Chen, S. H. Low and J. C. Doyle, On the dual scheduling algorithm: asymptotic optimality and throughput optimality, Submitted, 2006.
21. D. Bertsekas, *Nonlinear Programming*, 2nd edition, Athena Scientific, 1999.

22. N. Z. Shor, *Minimization Methods for Non-Differentiable Functions*, Springer-Verlag, 1985.
23. F. P. Kelly, A. K. Maulloo and D. K. H. Tan, Rate control for communication networks: Shadow prices, proportional fairness and stability, *Journal of Operations Research Society*, **49**(3):237-252, March 1998.
24. S. H. Low and D. E. Lapsley, Optimal flow control I: Basic algorithm and convergence, *IEEE/ACM Transactions on Networking*, **7**(6):861-874, December 1999.
25. A. Eryilmaz and R. Srikant, Fair resource allocation in wireless networks using queue-length-based scheduling and congestion control, *Proc. IEEE Infocom*, 2005. Submitted for journal publication.
26. A. Bar-Noy, A. Mayer, B. Schieber and M. Sudan, Guaranteeing fair service to persistent dependent tasks, *SIAM J. Comput.*, **27**(4):1168-1189, August 1998.
27. S. Asmussen, *Applied Probability and Queues*, 2nd edition, Springer, 2003.
28. M. Andrews, Instability of the proportional fair scheduling algorithm for HDR, *IEEE Transactions on Wireless Communications*, **3**(5):104-116, 2004.
29. T. Bonald and A. Proutiere, Wireless downlink data channels: user performance and cell dimensioning, *Proc. ACM Mobicom*, 2003.
30. G. Nong and M. Hamdi, On the provision of Quality of Service guarantees for input queued switches, *IEEE Communication Magazine*, December 2000.

8

The Combined Input and Crosspoint Queued Switch

Kenji Yoshigoe[1] and Ken Christensen[2]

[1] Dept. of Computer Science, University of Arkansas at Little Rock, Little Rock, AR 72204
[2] Dept. of Computer Science and Engineering, University of South Florida, Tampa, FL 33620

Input-queued (IQ) switches do not require the very high internal speedup that is necessary in output-queued (OQ) switches. IQ switches, however, require fairly complex scheduling of the switch matrix, need moderate internal speedup for stability, and cannot natively switch variable-length packets. The combined input and crosspoint queued (CICQ) switch combines input queueing (using virtual output queues) with a buffered crossbar. In this chapter, the history, performance, and future directions of the CICQ switch since its invention in 2000 are described. For a CICQ switch, switch matrix scheduling is simple, performance is high, and native switching of variable-length packets is possible. The CICQ switch also lends itself well to exploiting future higher VLSI densities. Future directions include a new concept of virtual crosspoint queueing to reduce the effects of internal feedback delays. Challenges remain in how to deliver QoS equivalent to that possible with OQ switches.

8.1 Introduction

Single-stage packet switches are the foundation of the Internet. The single-stage switch architecture can be used within LAN switches at the edge of a network and in routers in the core of a network. Many of these switches are also based on a crossbar switch fabric. The simplest architecture is output queued (OQ) where all buffering is at the output ports and the internal switch crossbar and all buffers must operate at a "speedup" equal to N times link speed (for an N port switch). This speedup requirement limits the scalability of OQ switches and thus a general interest in input-queued (IQ) switches has developed. IQ switches potentially require no speedup of buffer memory or of the crossbar switch fabric. However, IQ switches require methods to prevent head-of-line (HOL) blocking. Virtual output queueing (VOQ) first proposed by Tamir and Frazier in [1] and generally developed by Anderson *et al.* [2] and McKeown [3] eliminates HOL blocking at the input ports However, a VOQ switch requires non-trivial scheduling of the crossbar matrix (*i.e.* a maximal matching of input and output ports) to achieve high throughputs. Anderson *et al.* pioneered the first VOQ scheduling method with Parallel Iterative Matching (PIM) [2].

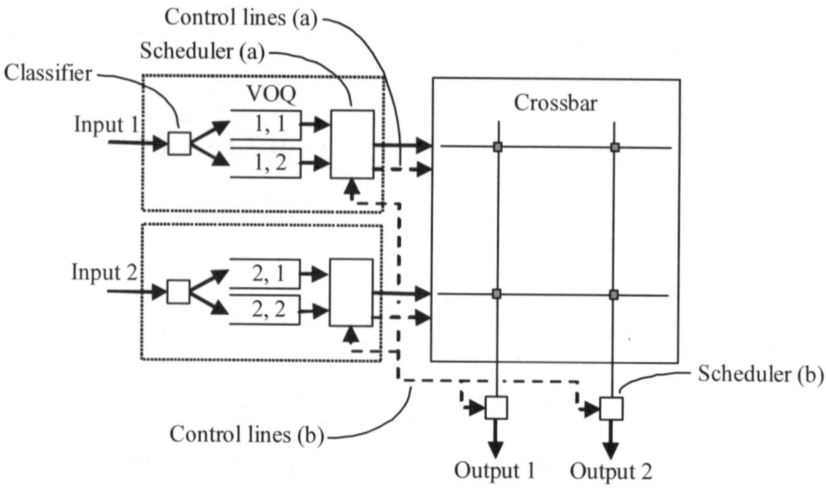

Figure 8.1. VOQ switch

McKeown improved upon PIM with iterative Slip, or iSLIP [3]. Many later works are variations on iSLIP that offer improvements either in lower delay or reduction in scheduler complexity. Figure 8.1 shows a two-port VOQ switch. The packet classifier determines which output port a packet is to be forwarded to and queues the packet in the associated VOQ buffer. The crossbar scheduler functionality is split between the input ports (a) and output ports (b). Control lines (a) from the input port schedulers control the crosspoints in the crossbar. Control lines (b) between the input and output ports are used to signal match requests and receive feedback on matches between the input and output ports. Inherent in all VOQ switches is the need to use fixed-length cells within the switch for synchronized scheduling of all ports. Thus, variable-length packets are segmented at the input port, switched as cells, and then reassembled into packets at the output port. This use of internal cells forces an almost $2\times$ speedup requirement for switching for worst case packets (these are packets of length $S + 1$ bytes for S byte cells). An additional $2\times$ speedup is needed for stability for 100% throughput for all possible inputs [4]. The complexity of VOQ switch scheduling is becoming a bottleneck for scalability to higher link speeds and/or greater port counts.

Many of the open problems existing in VOQ switches are eliminated by combining VOQ at the input ports with buffering in the crossbar. This combination – called a combined input and crosspoint-queued (CICQ) switch – was first proposed by Nabeshima [5] in 2000. This history of the CICQ switch is fully covered in Section 8.2. Figure 8.2 shows a three-port CICQ switch. Each input port has a classifier that queues an incoming packet into the VOQ matching the destination output port of the packet. Variable-length packets may be queued and forwarded directly (*i.e.* "native" packet switching where scheduling is not synchronized) or they may be segmented into fixed-length cells for synchronized scheduling. Both native packet and

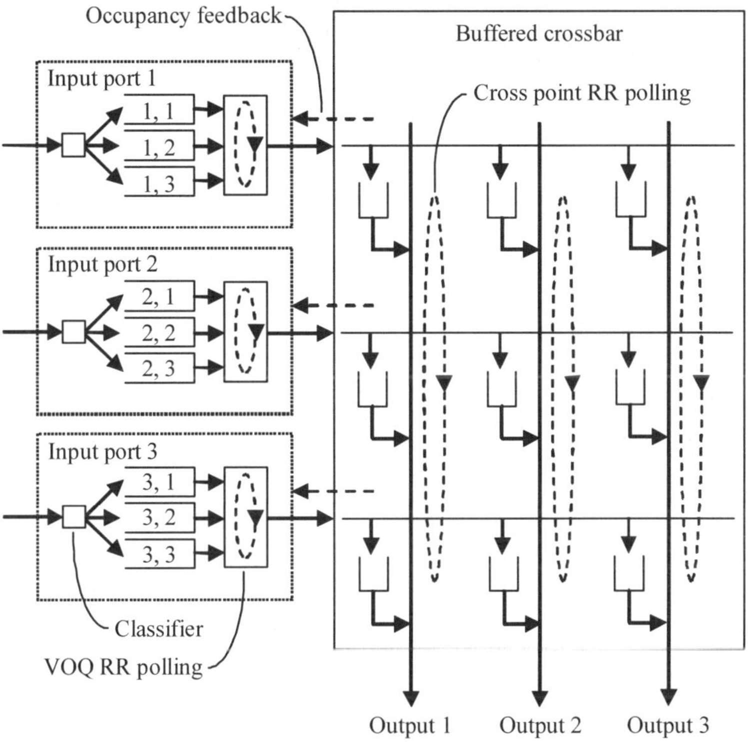

Figure 8.2. CICQ switch

cell-based switching are considered in this chapter. In the case of packet segmentation to cells, reassembly is performed at the output port. The input port includes a scheduler – shown in Figure 8.2 is a round-robin (RR) scheduler – to select the next eligible VOQ to forward its packet or cell to a crosspoint (CP) buffer. Occupancy feedback from the crosspoints is used to generate a mask of VOQs at each input port eligible to forward a packet or cell. If a CP buffer does not have sufficient available storage for a cell or packet, it masks-out its corresponding VOQ. The occupancy feedback can be thought of as a form of credit-based flow control. That is, a packet or cell cannot be forwarded to a CP buffer unless there is explicitly noted space available for it. Within the crossbar, scheduling of cell (or packet) forwarding from CP buffers to output ports is also based on RR arbitration in each "column" of the crossbar. The CICQ switch of Figure 8.2 is thus an RR/RR CICIQ switch and is the type of swtich studied in this chapter. Other types of scheduling are possible in the input ports and/or in the crossbar including Longest Queue First (LQF) and Oldest Cell First (OCF).

The CICQ switch addresses many of the open problems of IQ switches that use an unbuffered crossbar switch fabric. The open problems addressed include:

- Native switching of variable-length packets is possible. This eliminates the need for segmentation and reassembly logic. This may also reduce required internal speedup.
- A 2× speedup is not needed for inherent stability (*e.g.* to overcome insufficiency of maximal matching in IQ switches).
- Scheduling is simple and does not require coupling of input and output ports. CP buffer occupancy feedback from the CP buffers to the input ports is needed.
- The ever-increasing density of VLSI can be better exploited. Existing crossbar chips are often I/O bound leaving many cells or gates unused. Adding buffering and scheduling within the crossbar achieves better utilization of available gates in a crossbar chip.

One problem that is not addressed is that of QoS scheduling. IQ and CICQ switches are both deficient in this area compared to OQ switches. Existing Weighted Fair Queueing (WFQ) variants for QoS scheduling all require output buffering because all session information based on a single system virtual time must be taken into account to approximate General Processer Sharing (GPS).

8.2 History of the CICQ Switch

The concept of buffering in crossbars is almost 25 years old and is based on a 1982 patent by Bakka and Dieudonne [6]. In 1987, Nojima *et al.* implemented a bus matrix switch (BMX) using a large multi-cabinet design [7]. The BMX used a crossbar with CP buffers for packet queueing. The CP buffers were implemented with dual-port memories allowing asynchronous operation among input and output ports. Switching of variable length packets on each bus was possible. By increasing the number of buses, the BMX could increase the switch capacity. Simulation results confirmed that the switching delay was independent of the number of buses. A buffered crossbar switch is equivalent to an OQ switch that has dedicated memory for packets from each input. However, very large amounts of buffering are needed at each crosspoint, and this is not feasible.

This amount of buffer memory needed within the crossbar can be reduced by adding buffering at the input ports. Such a switch was investigated by Gupta *et al.* where look-ahead selection of a packet from an input queue was based on level of head-of-line (HOL) blocking [8]. A 16-port switch achieved 87.5% throughput for uniform traffic [8]. This work was extended (by Gupta *et al.*) to support two levels of delay-dependent priority classes, which resulted in an increase in throughput from 87.5% to 91% [9]. Switches with simple input queueing and a buffered crossbar were further studied by Doi and Yamanaka [10] and Re and Fantacci [11]. Re and Fantacci proved that the throughput for a CP buffered switch with FCFS queueing discipline and random selection policy can approach 100% throughput for uniform traffic. Stephens and Zhang proposed and modeled a buffered crossbar switch with input buffers that supports QoS and variable-length packets [12]. This switch had both buffering and packet fair queueing servers within the input ports, crossbars, and output ports. A speedup of slightly less than 2× was needed to support variable-length packets.

In 2000, Nabeshima was the first to propose combining VOQ and buffered crossbar [5]. He was also the first to use the term “CICQ” for a VOQ plus buffered crossbar switch. This terminology is followed in this chapter; that is, a CICQ switch contains VOQ and a buffered crossbar. In the switch proposed by Nabeshima, OCF is used where the HOL cell with the longest delay is selected for both the input port (by a polling of all VOQs) and the CP buffer (by a polling of all CP buffers). The CICQ switch entirely eliminates HOL blocking, and it has a lower mean cell delay than a pure IQ switch above 65% loads. More importantly, the VOQ CICQ switch significantly reduced the amount of buffering required in the crossbar. As VLSI density has continued to increase, it is now feasible to implement small amounts of buffering at each crosspoint in a crossbar. Xilinx implements a buffered crossbar in FPGA technology for its Virtex-Extended Memory (Virtex-EM) devices [13]. In 2005, the commercially available Xilinx XC2VP100 device contained 1 Mbyte of select RAM, enabling 4 Kbytes of buffer space for each crosspoint of a 16-port switch. For a cell-based switch, buffering equal to one or two cells at each crosspoint is sufficient. The CP buffer occupancy status is reported from each crossbar row to its input port where an independent RR polled selection of VOQs is made for the next available CP. The buffer occupancy status must be reported at a rate equal to the maximum cell transmission rate, which can be done asynchronously for each input port. No communication of state is necessary between output and input ports.

Various scheduling algorithms for the CICQ switch have been investigated in the past 5 years. A CICQ switch can have RR polling of the VOQs at the input ports and RR polling of the CP buffers, and is thus an RR/RR CICQ switch [14]. The RR/RR CICQ switch natively supports variable-length packets. Yoshigoe and Christensen showed that the CICQ switch has a lower delay than a VOQ IQ switch with iSLIP for both cell and packet switching under uniform traffic [14]. A Combined Input-One-Cell-CP Buffer crossbar (CIXB 1) with VOQs at the inputs and RR arbitration (identical to RR/RR CICQ) was independently proposed at the same time by Rojas-Cessa *et al.* [15]. This switch design was shown to achieve 100% throughput under uniform traffic [15]. The mean delay of the CIXB-1 switch was proportional to burst length and very close to that of an OQ switch. A Combined Input-CP-Output Buffered (CIXOB-k, where k is the size of the CP buffer) with VOQs at the input ports and RR arbitration requires buffers at each input, output, and CP [16]. A CIXOB-k switch could achieve 100% throughput under uniform as well as non-uniform traffic. A full-scale system design of a terabit switch incorporating ideas for the CIXOB-k switch architecture is described by Chao [17]. Performance degradation due to synchronization effects of multiple input ports transmitting cells to a same destination output was studied by Han *et al.* [18, 19]. To reduce the input synchronization, the starting points of the RR arbiters were incremented by one independent of a matching result. This reduced the mean delay for bursty traffic [18,19]. Different combinations of RR, LQF, and OCF at input and output scheduler were evaluated. Simulation results showed that there was only a slight difference in mean delay for both uniform Bernoulli and Interrupted Bernoulli Process (IBP) traffic [20].

A major issue in switch matrix scheduling is stability. A stable switch has bounded queue length for all possible schedulable loads. Both weight-based [21] and priority-based [22] schemes have been shown to provide 100% throughput for any admissible flow. CICQ switches can achieve stability if OCF or LQF is used to select VOQs in an input port [21]; however, both OCF and LQF require comparisons between all N ports during each scheduling cycle. This requires either N sequential comparisons or $\log_2(N)$ comparisons with a tree circuit containing $N-1$ comparators. A method first proposed by Yoshigoe *et al.* in [23] and further investigated in [24] does not require $\log_2(N)$ comparisons to achieve stability. In the proposed method, when a VOQ in an input port is selected to transfer a cell in the next cycle, a threshold comparison is made. As long as the current VOQ queue length exceeds a set *THRESHOLD*, then up to *BURST* cells can be transmitted from the VOQ before another VOQ from the same input port is allowed to be matched. An RR-based arbitration scheme with adaptable-size frame was proposed by Rojas-Cessa in [25] and further evaluated in [26]. This scheme can adapt to frame size based on the amount of service that both VOQs and CP buffers receive. A similar scheme was proposed by Rojas-Cessa where frame size is determined by the cell occupancy of the VOQ at the time of frame service completion [27]. A scheduling algorithm based on the Shortest internal Buffer First (SBF) at the input side and the Longest internal Buffer First (LBF) at the output side has lower mean delay than OCF/OCF CICQ [5] and LQF/RR CICQ [21] switches under bursty uniform traffic as well as under unbalanced traffic while reducing its hardware and timing complexity [28]. A similar approach by Zhang and Bhuyan in [29], using a Shortest Crosspoint Buffer First (SCBF) algorithm for input arbiters with any work-conserving output arbiters, was shown to achieve 100% throughput for any admissible traffic. Analysis by Lin and McKeown showed that the throughput of a buffered crossbar switch increases with switch size, which contrasts with unbuffered crossbar switches [30].

Only recently has QoS been addressed in the context of CICQ switch architectures. In [31] Motoyama and Arantes tagged arriving cells at the input port with their arrival time and forwarded the tagged cells to the CP buffers. The amount of bandwidth needed to allocate to a flow in the switch to guarantee delay was determined by Duan and Daigle [32]. WFQ for a CICQ switch was implemented and its fairness properties were studied by Chrysos and Katevenis [33]. Magill *et al.* [34] and Mhamdi and Hamdi [35] emulated an OQ switch with a CICQ switch with $2\times$ speedup. Chuang *et al.* [36] and Giaconne *et al.* [37] proved that a $2\times$ speedup can guarantee 100% throughput in a CICQ switch. Chuang *et al.* also proved that a $2\times$ speedup is sufficient for rate guarantees and a $3\times$ speedup is sufficient for delay guarantees [36]. The smooth multiplexer of He *et al.* emphasized QoS over work-conservation [38]. The smoothed multiplexer used rate-based flow control where the weight of each flow is considered for scheduling. The study proved that the occupancy of each crosspoint buffer in the proposed smoothed CICQ switch never exceeded four cells regardless of internal switch fabric latency and line rate. Further investigation is needed into QoS in the CICQ switch architecture – in particular how to feasibly implement QoS scheduling.

An RR/RR CICQ switch can natively forward variable-length packets, and such a switch requires no speedup to compensate for padding bytes nor is segmentation/reassembly of the packets at the input/output ports needed. However, it is unfair [39]. To eliminate this unfairness, a block transfer mechanism that transfers up to a predefined number of bytes of packet data from a selected VOQ was proposed by Yoshigoe and Christensen [23] and Yoshigoe *et al.* [39]. The transfer block can consist of multiples of partial and/or entire packet. A similar approach was investigated by Katevenis and Passas [40].

To accommodate higher link speeds and greater port counts, switches are being implemented in multiple, distributed units. These units may be several meters apart introducing significant propagation delay between the line cards and switch fabric. An IBM research report by Abel *et al.* was the first work that describes the implementation of the distributed CICQ switch where line cards and switch fabrics are distributed in multiple units [41]. Since then, several studies have addressed the issue of CP buffer size related to a large RTT delay (here RTT is measured in cell times) between input ports and CP buffer. Luijten *et al.* investigated a priority elevation mechanism [42]. Temporary elevation of lower priority packets when they are already in the switch fabric was used to reduce blocking of higher-priority packets. The CP buffer size was significantly reduced from N^2PT_{RTT} to $N^2T_{RTT} + P - 1$ where T_{RTT} is the RTT. A two-lane buffered crossbar design was proposed to handle more than two levels of priority traffic using only two queues per CP by Chrysos [43] and Chrysos and Katevenis [44]. Gramsamer *et al.* showed that the CICQ switch with a CP buffer size that can hold 60% of back-to-back cells in transit between the line card and the CP buffer had an acceptable performance for bursty traffic [45]. Simulation evaluation of the effect of RTT and CP buffer size for variable-length packets has been carried-out by Katevenis *et al.* [46]. As link data rates and internal cable lengths increase, the minimum number of feedback credits needed to maintain work conservation of the switch increases. Consequently, the switch fabric will no longer be able to implement CP buffers sufficient to maintain work conservation of the switch. A load-balanced CICQ switch was proposed by Rojas-Cessa *et al.* [47]. An extra switch stage that serves as a load-balancer was inserted between the input ports and buffered crossbar to relax the CP buffer size such that flows with high data rates can be handled when CP buffer size is smaller than RTT. The CP buffer size was reduced by a factor of N independent of the RTT value. However, there is an additional cost for implementing a load-balancer. CICQ switches that scale independently of the growth of the RTT value are needed. This remains an open research challenge. Section 8.6 describes one possible new direction.

8.3 Performance of CICQ Cell Switching

The primary performance measure of interest in a packet or cell switch is switch delay. This is the delay measured from the first bit of a packet or cell entering the switch at an input port, to the first bit of the packet or cell departing from an output port. It is desirable to minimize mean switch delay and also minimize the variability of this delay.

In this section, the cell switching performance of an RR/RR CICQ switch is evaluated. In the next section, performance is evaluated for native packet switching. Performance evaluation was done using discrete-event simulation models of VOQ IQ (using iSLIP scheduling), OQ, and CICQ switch architectures. The models were built using the commercial CSIM simulation library [48] and validated against published results [49–51]. All simulation experiments were run using batch means convergence until a 2% accuracy level was achieved with a 95% confidence interval, unless otherwise stated. The simulation models are freely available from the authors of this chapter [52].

8.3.1 Traffic Models

To evaluate cell switching performance, Bernoulli and Interrupted Bernoulli Process (IBP) arrivals were modeled for a slotted (cell) system. The Bernoulli model is a "classic" traffic model for evaluating switch performance [2, 3, 21, 49, 53]. IBP arrivals are frequently used to approximate the bursty nature of packet switched traffic [14, 20, 54]. The IBP is a two-state traffic model alternating between busy (ON) and idle (OFF). For Bernoulli arrivals at rate λ ($0 < \lambda < 1$) the probability of a slot having an arriving cell is λ and being empty is $1 - \lambda$. The arrival rate (λ) is the offered load, ρ. For IBP arrivals,

$$\alpha = \Pr\left[\text{arrival at } t \mid \text{IBP is in ON state}\right]$$

$$p = \Pr\left[\text{IBP is in ON state at} t + 1 \mid \text{IBP is in ON state at } t\right], \text{ and}$$

$$q = Pr\left[\text{IBP is in OFF state at } t + 1 | \text{ IBP is in OFF state at } t\right]$$

The mean length of an ON period is $1/\left(1 - p\right)$ and the mean length of an OFF period is $1/\left(1 - q\right)$. The OFF period is at least one slot. The offered load is

$$\rho = \frac{\alpha(1 - q)}{2 - p - q} \tag{8.1}$$

The Coefficient of Variation (CoV) for IBP arrival is given by

$$CoV = 1 + \alpha\left(\frac{\left(1 - p\right)\left(p + q\right)}{\left(2 - p - q\right)^2} - 1\right) \tag{8.2}$$

In the experiments in this chapter, α was always set to 1 so that traffic was generated at full data rate in the ON state. The parameters p and q were varied to achieve a desired CoV and offered load.

8.3.2 Simulation Experiments

For all simulation experiments, the switch delay was measured in cell times. VOQ buffer size was assumed to be infinite in all cases. For the CICQ switch model, each CQ buffer size was equal to one cell length (64 bytes). For the iSLIP IQ switch model, four iterations of the iSLIP algorithm were implemented for each scheduling cycle. The switch was modeled with 16 input ports and 16 output ports ($N = 16$). The experiments were:

Bernoulli experiment (balanced): Bernoulli arrival of cells with uniformly selected outputs. Offered load ranged from 50% to 98%. This experiment evaluated switch performance for smooth and balanced traffic.

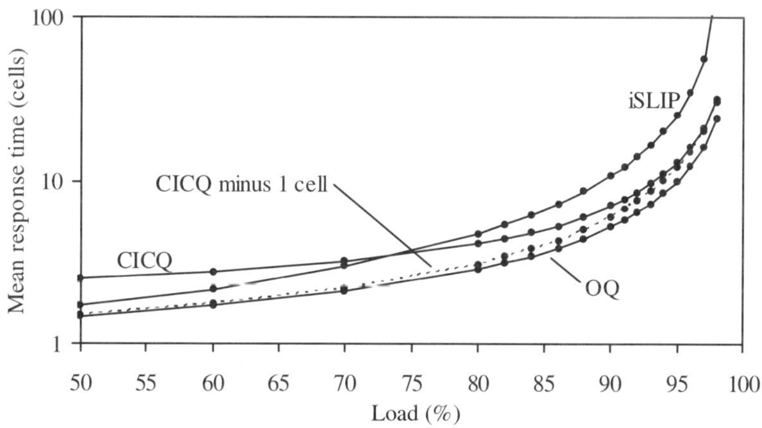

Figure 8.3. Mean response time (balanced Bernoulli arrivals)

Bernoulli experiment (unbalanced): Bernoulli arrival of cells where half of the input ports were the source of exactly four flows and half were the source of exactly eight flows. Each flow rate was identical. Thus, half the input ports were loaded at λ and the other half were loaded at 2λ. Eight random permutations were used to define the flows. All eight flows defined by the permutations were used for the heavily loaded input ports, and only the first four permutations were used to define the flows for lightly loaded input ports. Each output port supported at least four flows and at most eight flows. The offered load ranged from 50% to 98%. This experiment evaluated switch performance for smooth and unbalanced traffic.

IBP experiment: IBP arrivals of cells with uniformly selected outputs for bursts (*i.e.* for ON periods) of cells. The CoV was fixed at 2.0 and the p and q values solved (using Equations (8.1) and (8.2)) for offered loads from 50% to 90%. This experiment evaluated switch performance for bursty traffic.

Figures 8.3, 8.4 and 8.5 show the mean response time (switch delay) results for the three experiments. For the Bernoulli experiment with balanced selection of output ports, the mean response time for the CICQ switch was lower than that of iSLIP for offered loads greater than 75%. At lower offered loads, the two store-and-forward

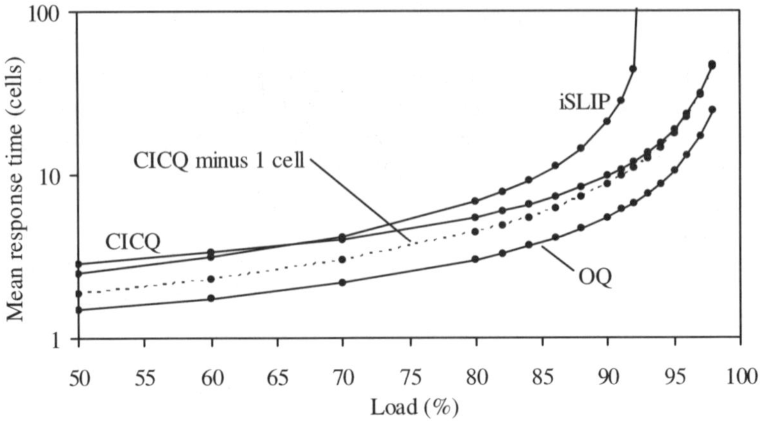

Figure 8.4. Mean response time (unbalanced Bernoulli arrivals)

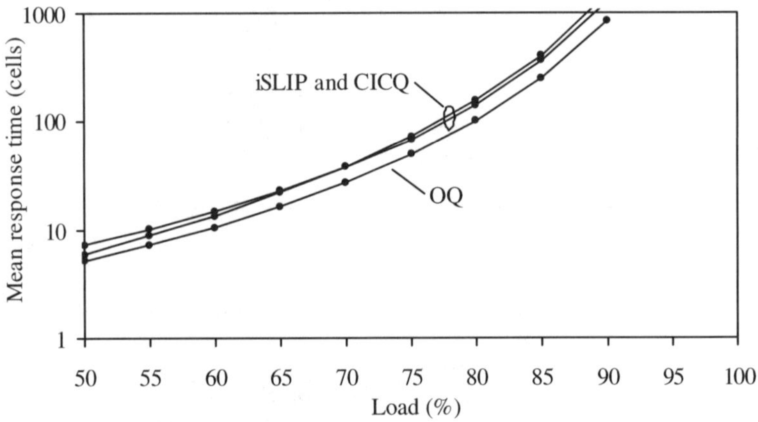

Figure 8.5. Mean response time (balanced IBP arrivals)

operations within the CICQ switch dominate the response time (the crossbar buffering was implemented as store-and-forward memory in the simulation model). If one store-and-forward delay is removed, the results shown with the dotted line in Figure 8.3 are achieved and the response time is less than that of an iSLIP switch for all offered loads. The OQ switch remains as a lower bound for delay. For the Bernoulli experiment with unbalanced arrivals, the iSLIP switch becomes unstable above 92% offered loads where the CICQ switch does not become unstable. Speedup is needed in the iSLIP switch to overcome this inherent instability. For the IBP experiment, the iSLIP switch had slightly lower delay than the CICQ switch at low offered load while they had roughly similar delay at high load. Again, the OQ switch remains as a lower bound for delay. Bursty traffic increases switch delay and reduces the difference in performance between the different switch architectures.

8.4 Performance of CICQ Packet Switching

In this section, the performance of an RR/RR CICQ switch is evaluated for native packet switching. As for the cell switching evaluation, all simulation experiments were run using a batch means convergence until a 2% accuracy level was achieved with a 95% confidence interval, unless otherwise stated. The simulation models are freely available from the authors of this chapter [52]. The models require the CSIM library [48].

8.4.1 Traffic Models

To evaluate packet switching performance, packet lengths were independently pulled from an empirical "USF distribution" based on over 5 million packets collected at about noon during a week day in November 2001 from the University of South Florida Gigabit Ethernet backbone. Figure 8.6 shows a histogram of the measured packet lengths. The mean packet length was 364.7 bytes. The most common packet length was 64 bytes (with 41.5%) followed by 1518 bytes (8.2%), 558 bytes (7.0%), 90 bytes (5.9%), and 570 bytes (5.5%). All other packet lengths occurred at less than 2.5% frequency.

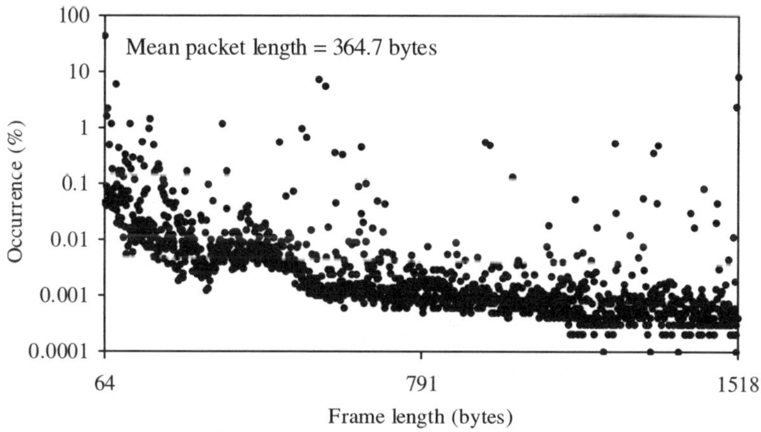

Figure 8.6. USF distribution of packet lengths

8.4.2 Simulation Experiments

For all simulation experiments the mean switch delay was measured in microseconds. VOQ buffer size was assumed to be infinite in all cases. For the CICQ switch model, each CQ buffer size was equal to 1518 bytes (*i.e.* maximum Ethernet packet length). For the iSLIP IQ model, four iterations of the iSLIP algorithm were implemented using Marsan *et al.* cell trains [55]. In the cell train modification of iSLIP, the

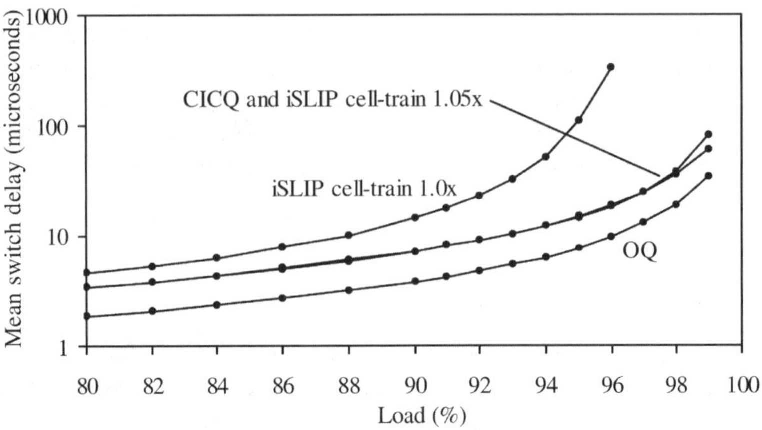

Figure 8.7. Mean response time (balanced packets)

matching of a head-of-packet cell was maintained until all cells of that packet were forwarded. The use of cell trains provides better performance than individual cell-by-cell scheduling for variable-length packets for iSLIP. With a mean packet length of 364.7 bytes, six 64-byte cells of total 384 bytes are needed, which is a 5% overhead. Thus, for the iSLIP IQ switch an internal speedup of 1.05× of the buffer memory and crossbar was modeled. The switch was modeled with 16 ports ($N = 16$). The experiments were:

Packet experiment (balanced): Poisson arrivals of variable length USF distributed packets with uniformly selected outputs. The offered load was varied from 80% to 98%.

Packet experiment (unbalanced): Poisson arrivals of variable length USF distributed packets with unbalanced load (same as for the cell switching Bernoulli unbalanced experiment). Offered load was varied from 80% to 98%.

Figures 8.7 and 8.8 show the mean response time (switch delay) results for the packet experiments. For balanced traffic, the iSLIP cell train without speedup could not carry an offered load above 96%. With 1.05× speedup, stability was achieved. The mean delays of the iSLIP with cell train and 1.05× speedup, and that of the CICQ switch are similar. The speedup used with the iSLIP with cell train compensates for the transferring of empty bytes, achieving stability for all offered loads measured. The CICQ switch natively supports variable-length packets. As was expected, the OQ switch had the lowest mean delay. For unbalanced traffic, the iSLIP with cell train, the iSLIP with cell train and 1.05× speedup, and the CICQ switch become unstable above 92%, 96%, and 98% load, respectively. The OQ switch is stable for all offered loads measured.

CICQ switches can natively forward variable-length packets and do not require speedup due to transfers of padding bytes. However, a CICQ switch forwarding

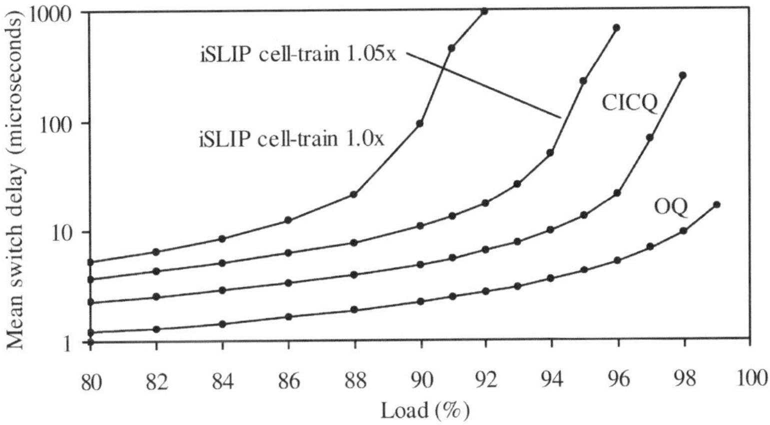

Figure 8.8. Mean response time (unbalanced packets)

native variable-length packets is not work conserving when the transfer of a large packet from VOQ_{ij} to CP_{ij} "blocks" transfer of a small packet from VOQ_{ik} ($k \neq j$). For a simple scenario, consider a two-port CICQ switch where CP_{12} and CP_{22} have a packet of size L buffered. If both VOQ_{11} and VOQ_{21} were selected for transferring packets of size greater than $2L$, output link 2 entirely drains the packets from CP_{12} and CP_{22} before the transfer of packets to CP_{11} and CP_{21} are completed. Thus, both CP_{12} and CP_{22} become empty, and output link 2 temporarily becomes idle. Output link 2 would not have been idle if either VOQ_{12} or VOQ_{22} were selected. This condition can exist in a switch with any port count. This lack of work conservation can be addressed with internal speedup or by a block transfer mechanism [23, 39] or some variant of block transfer [40].

8.5 Design of Fast Round-robin Arbiters

Many switch architectures, including iSLIP [3], use round-robin (RR) arbiters as part of their switch matrix scheduling. The RR/RR CICQ switch uses two levels of RR arbitration. The RR arbiter is the bottleneck in a CICQ switch as the number of ports and/or link data rates increase. For example, for a CICQ switch with 16 ports and a 100 Gb/s link data rate a 64-byte cell must be forwarded in 5.12 nanoseconds, and a worst case scheduling cycle also completed in this time (*i.e.* with 0.32 nanoseconds per input port polled). The worst case scheduling cycle is N ports (*e.g.* one port has packets or cells queued and $N - 1$ empty ports must be polled between each non-empty poll).

For an RR/RR CICQ switch to achieve 100% throughput, the RR arbiters must be work conserving. Consider a slotted system with N queues, Q_i, where $i = 1, 2, \ldots, N$. Each queue buffers fixed-length cells arriving from external sources. The

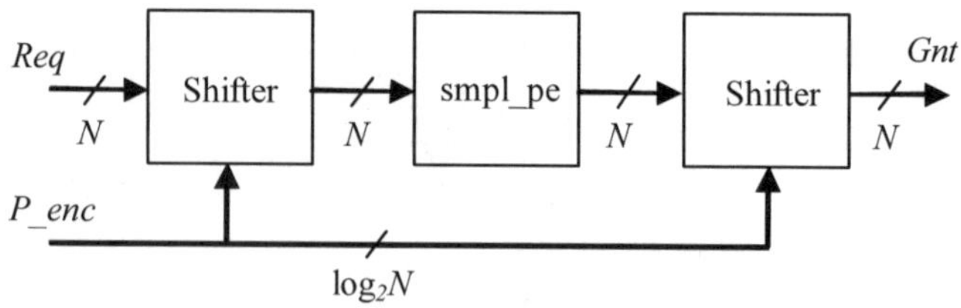

Figure 8.9. Shift encoder design

cell queues may correspond to VOQs at a switch input port. The time to forward a cell is T_c. A poller visits queues Q_i in sequential RR fashion with a delay of T_p for each queue visited. The delay T_p occurs whether the visited queue is occupied or empty. If $N \cdot T_p < T_c$ then a simple two-stage RR arbiter that can select the next queue while the currently selected queue is forwarding a cell is sufficient (the arbiter is work-conserving) such that the output link is never idle if there are cells queued in any of the N queues. Most existing RR arbiters are based on a two-stage approach in which scheduling is done simultaneously with cell transmission. A cell forwarded in time slot t_i is scheduled in a previous time slot t_{i-1}. In existing two-stage switch designs that select one cell per scheduling cycle, 100% throughput for the single cell selected per poll can only be achieved if $N \cdot T_p < T_c$. If $N \cdot T_p > T_c$, 100% throughput can be achieved only if more than one cell can be selected and forwarded per poll. A scheduler may exhibit fairness in the short term and must exhibit fairness in the long term. In a short-term fair scheduler, all N nodes receive an opportunity to forward a cell within every N cell forwarding time. In a long-term fair scheduler, all N nodes receive an opportunity to forward a cell in a finite time (*i.e.* scheduling delay is bounded). In this section, designs for fast RR arbiters for the two cases, (1) short-term fair where $N \cdot T_p < T_c$ and (2) long-term fair where $N \cdot T_p > T_c$, are investigated.

8.5.1 Existing RR Arbiter Designs

Previous work has been done in designing fast RR arbiters at both the transistor and gate level. A priority encoder was implemented by Delgado-Frias and Nyathi at the CMOS transistor level [56]. This implementation used a priority look-ahead approach similar to that of a look-ahead adder. The design (implemented in 1 micron CMOS) achieved a priority encoding in 4.1 nanoseconds for a 32-bit input.

A well-known gate-level design for a short-term fair RR arbiter is the double barrel-shift RR poller (called SHFT_ENC in [57]), which is shown in Figure 8.9. It consists of two barrel shifters and a simple priority encoder, smpl_pe. Request bits Req of size N are rotated by an amount P_enc (P_enc is $\log_2(N)$ bits) to indicate the queue with the currently selected buffer. This is then input into another smpl_pe and again rotated by P_enc in the reverse direction. The outputs are grant bits Gnt of size N and a bit, $anyGnt$, indicating whether there is a grant. The barrel shifters dominate the critical path delay.

Existing short-term fair RR arbiters can be categorized as sequential polling, non-sequential polling, tree arbitration, and pipeline structures. Sequential polling is the simplest implementation of RR polling, and it has $O(N)$ scalability with each new node adding a delay T_p to the scheduling delay. An example of sequential polling is fully serial ripple (RIPPLE) described in [57]. Token tunneling, a form of non-sequential polling, proposed by Chao, allows a pointer to skip a set of ports if none of the ports has packets to send [17]. Token tunneling reduces the arbitration time of sequential polling to $O(\sqrt{N})$.

Tree-structured RR arbiter designs [57–60] reduce the arbitration time over a sequential design. The arbitration time of the ping-pong arbitration (PPA) scheme by Chao *et al.* for an N-input switch is proportional to $\log_2 \lceil N/2 \rceil$ [58]. The arbitration time is only 11 gate delays for a 256-port switch. The parallel RR arbiter (PRRA) by Zheng *et al.* uses a binary tree structure [60]. Similar to the PPA design, the PRRA has $O(\log_2(N))$ gate delays but it resolves an unfairness problem in the PPA scheduler. Arbitration time of the tree-structured RR arbiters described in [57, 59] is also $O(\log_2(N))$. The Exhaustive (EXH) in [57] uses N duplicate copies of a simple priority encoder, and the selection of priority encoder is made by the programmable priority input. Thus, this design does not scale for a large N. The PROPOSED design in [57] is a look-ahead approach that eliminates the programmable part of a programmable priority encoder (PPE) by pre-processing inputs. PROPOSED eliminates a combinational feedback loop that is difficult for synthesis tools to optimize and eliminates a long critical path from a programmable highest priority level.

Pipelined arbiter designs are used in several switch architectures including those of [61–63]. Round-robin greedy scheduling (RRGS) can scale to a large switch because the amount of information transferred among function modules is small [63]. However, the scheduling delay increases in proportion to the number of switch ports, and scheduling can be unfair. The group-pipeline scheduler (GPS) by Motoki *et al.* improves on RRGS by dividing N nodes into K groups (N/K nodes per a group), and assigns an RRGS function module to each group [62]. It has a smaller arbitration time and better fairness than RRGS. An FPGA implementation of RR scheduling using a pipelined priority encoder and barrel shifter was developed by Huajin *et al.* [61]. Encoded bits are divided and input to multiple smaller priority encoder units. Outputs from these units are merged and input to another priority encoder to obtain the final encoded result. An arbitration time of $O(\log_2(N))$ can be obtained at the cost of using more FPGA space.

Some pipelined arbiters are not of a two-stage design. One example is the pipeline-based concurrent round-robin dispatching scheme using multiple subschedulers by Oki *et al.* [64]. Each subscheduler provides a dispatching result in every Pth scheduling cycle given P subschedulers.

8.5.2 A New Short-term Fair RR Arbiter – The Masked Priority Encoder (MPE)

A new, fast short-term fair masked priority encoder (MPE) poller design was first developed and evaluated by Yoshigoe *et al.* [65]. The MPE is a priority encoder that

1. Generate masking bits *Msk* based on the previously selected VOQ value.
2. Mask-out request bits that are equal or less than the value of the previously selected VOQ.
3. If masked requests *M_req* is non-zero then select *M_req*. Otherwise select *Req*.
4. Encode the selected bits (*M_req* or *Req*) with smpl_pe.

Figure 8.10. MPE RR arbiter algorithm

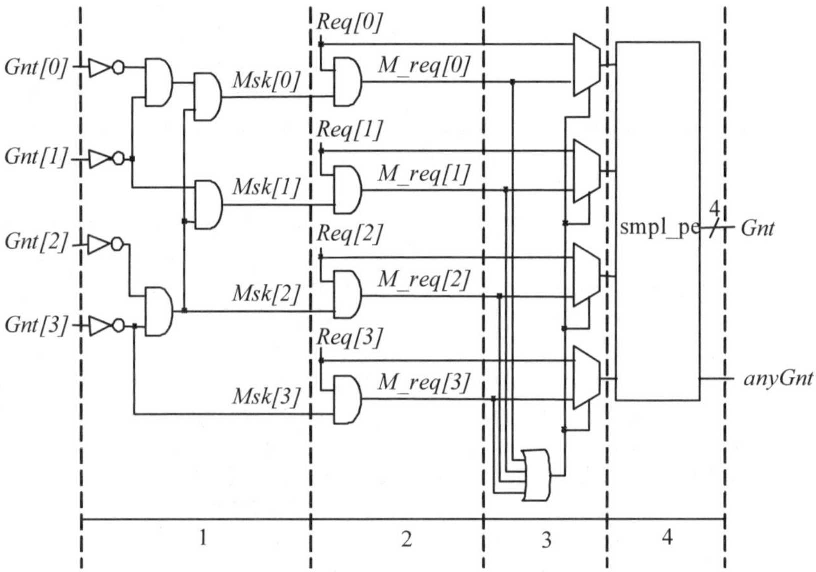

Figure 8.11. MPE RR poller design

uses bit-wise masking to select an appropriate VOQ. As with the PROPOSED design [57], no programmability is required, and only a smpl_pe is needed. The four basic steps in the MPE are shown in Figure 8.10. The masking bits are generated by $Msk[i] = \overline{Gnt[i]} \cdot \overline{Gnt[i+1]} \cdot \overline{Gnt[i+2]} \cdot \ldots \cdot \overline{Gnt[N-1]}$. The MPE directly uses a previously derived N-bit grant value for the next polling. Thus, it requires neither an encoder or decoder to convert the N-bit form to or from a $\log_2(N)$-bit form. Figure 8.11 shows the logic diagram of the MPE. To evaluate the performance of the MPE, the programmable priority encoder RR poller designs in [57] were implemented using VHDL, and simulated with the Xilinx WebPACK 4.2 ModelSim XE [66]. The targeted device was the Xilinx Virtex II XC2V40 FG256. Simulations were run with the same time and space optimization settings for all designs. The delay and space requirement were measured for each design. Table 8.1 shows the delay (in nanoseconds) and Table 8.2 shows the space in basic elements (BEL) of the RR arbiter designs. The last row in the tables indicates the improvement of MPE over the design with the next best performance. BELs are the building blocks that make up a component configurable logic block for an FPGA and include function generators, flip-flops, carry logic, and RAM. The relative results do not exactly match with the results in [57]. In [57] two-input gate equivalents were used to size the designs. The design in [57] uses a digital signal processor (DSP) as the target device; however, the target device used for this study was an FPGA, which is capable of handling the high-speed data on the chip. This difference in targeted devices also results in the use of different simulation tools and configurations.

Table 8.1. Evaluation of delay (nanoseconds)

Design	$N = 8$	$N = 16$	$N = 32$	$N = 64$
RIPPLE	17	24	41	73
CLA	14	17	23	23
EXH	10	16	26	50
SHFT_ENC	15	24	37	64
PROPOSED	13	21	33	55
MPE	10	11	13	16
Improvement	0.0 %	47.6 %	43.5 %	30.4 %

Table 8.2. Evaluation of space (FPGA BELs)

Design	$N = 8$	$N = 16$	$N = 32$	$N = 64$
RIPPLE	17	31	126	380
CLA	21	41	145	418
EXH	132	473	2391	10134
SHFT_ENC	58	143	350	836
PROPOSED	37	74	150	318
MPE	65	134	355	798
Improvement	–282.6%	–332.3 %	–181.7 %	–150.9 %

The results show that the MPE had lower delay than any other design for all measured values of N. However, it required more space than any other design, except EXH. For modern VLSI, space is rarely the constraining factor. The better delay performance of the MPE is due to the fact that the MPE uses N bits to determine the value for the next poll. The MPE does not require an encoder or decoder to convert the N-bit form to and from a $\log_2(N)$-bit form, which would result in a speedup at the cost of space required to accommodate N bits versus $\log_2(N)$ bits.

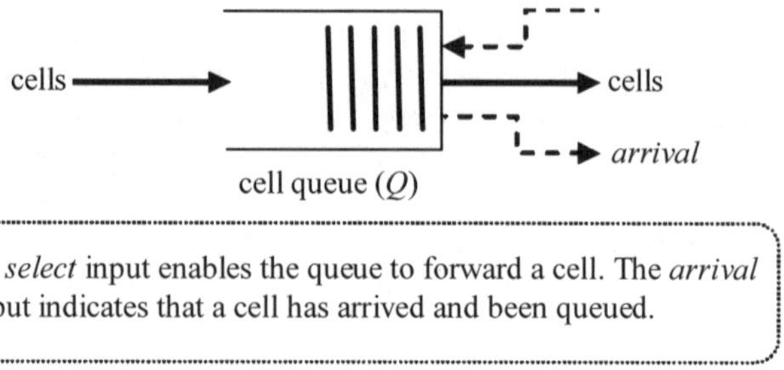

Figure 8.12. Queue with control and data lines

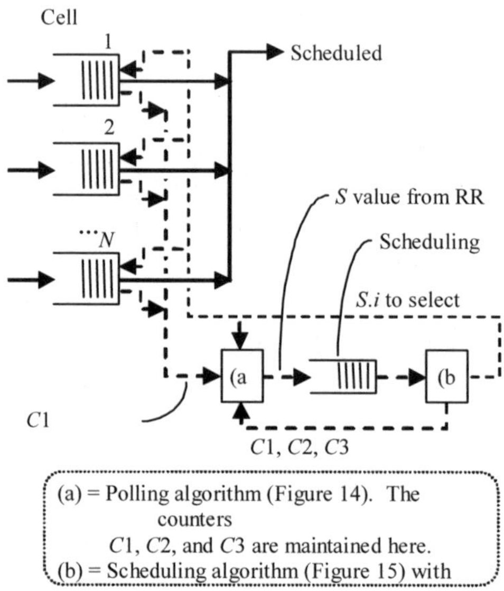

Figure 8.13. Cell queues and scheduling queue

8.5.3 A New Fast Long-term Fair RR Arbiter – the Overlapped RR (ORR) Arbiter

To improve the scalability of RR polling, a long-term fair overlapped round-robin (ORR) arbiter that fully overlaps polling and cell scheduling was proposed by Yoshigoe and Christensen in [67]. In a system of N queues, each queue has one control input (*select*) and one control output (*arrival*). Figure 8.12 shows a cell queue and Figure 8.13 shows the system of N queues with an (a) RR polling unit and (b) cell scheduling unit. The polling algorithm is shown in Figure 8.14 and the scheduling algorithm in Figure 8.15. A counter $C1_i$ is incremented on cell arrivals to Q_i and decremented on scheduled cell departures. The *arrival* output causes the increment of $C1_i$. The decrementing of $C1_i$ is caused by the scheduling algorithm. The counter

```
1. do forever
2.     i = mod(i, N) + 1
3.     while (C3 > K) wait
4.     mark = min(K, C1_i – C2_i)
5.     if (mark > 0)
6.         C2_i = C2_i + mark
7.         C3 = C3 + mark
8.         S.i = i
9.         S.m = mark
10.        queue S to the scheduling queue
```

Figure 8.14. Polling algorithm

```
do forever
    if (the scheduling queue is non-empty)
        S = dequeue from scheduling queue
        set select for queue S.i
        do for j = 1 to S.m
            wait for a cell to finish forwarding
            decrement C1_S.i, C2_S.i, and C3
        reset select for queue number S.i
```

Figure 8.15. Scheduling algorithm

$C1_i$ represents the number of cells currently queued in queue Q_i. A counter $C2_i$ is decremented on cell departures from Q_i and is increased in the polling algorithm shown in Figure 8.14. The counter $C2_i$ represents the number of cells in a queue "marked" for forwarding. At all times, $C1_i - C2_i \geq 0$.The *select* input line is used to select a queue for forwarding cells. Only one *select* line can be active for any given cell slot. A single counter $C3$ representing the number of cells permitted to be forwarded in the scheduling queue is also maintained. All counters are stored in the polling unit. A constant value, K, is used in the polling unit. The counter $C3$ and the setting of K are described later. The polling algorithm (Figure 8.14) "visits" each queue by testing whether $C1_i - C2_i > 0$. (line 4). If this holds, then there are unmarked cells in the queue. When a queue is visited and the *mark* value (line 4) is non-zero, the counters $C2_i$ and $C3$ are updated and a scheduling value, S, comprising the queue index, i, concatenated with the number of cells marked in this visit, m $(1 \leq m \leq K)$, is queued in a special scheduling queue. The polling time T_p is incurred in lines 2 to 10 of the polling algorithm and in the time to increment $C1_i$ For this study, the notation $S.i$ and $S.m$ are used to mean the index value and marked cell count, respectively, for a given value of S. The value S is of size $\log_2(N)+\log_2(K)$ bits. Line 3 in the polling algorithm stops the polling if the value of $C3$ exceeds K. The counter $C3$ contains the sum of $S.m$ currently queued in the service queue. The poll stopping in line 3 is essential to improving long-term fairness, and its properties are discussed later in this section. The scheduling algorithm (Figure 8.15) dequeues S from the scheduling queue when all currently scheduled cells have been forwarded. For example, if the currently dequeued scheduling value has $S.m$ equal to 3, then after 3 cell forwarding times, the next queued scheduling value will be dequeued. The index $S.i$ is the queue to be issued a select for forwarding of $S.m$ cells. The polling and scheduling algorithms run concurrently. The value of K is set so that work conservation is achieved for all possible cases of queued cells in the N queues. The value of K also binds the maximum delay a cell arriving to an empty queue will experience (from [67]):

Lemma 1. *The smallest integer K needed for the ORR scheduling to achieve work conservation for all cases of queued cells in the N queues can be derived as*

$$K = \left\lceil N \frac{T_p}{T_c} \right\rceil$$

Proof. A time to poll all N nodes; NT_p, divided by T_c is the total cell forwarding time required to poll all N nodes. That is, NT_p/T_c cells are forwarded during one RR scheduling cycle. If this RR scheduling time is less than the cell forwarding rate, at least one cell is scheduled during one cell forwarding time. Thus the system becomes work-conserving for any $K > NT_p/T_c$. The least integer greater than or equal to NT_p/T_c is a ceiling of NT_p/T_c.

Theorem 1. *A new HOL cell at any queue of the ORR arbiter can be forwarded in less than $K \cdot (N-1) + 2K + 1$ cell forwarding times.*

Proof. By definition, the ORR poller visits any of N queues in every N polling time where each queue marks up to K cells per polling time. Thus, a HOL cell at any queue has to wait, at most, $K(N-1)$ cell forwarding times if the scheduled queue was empty. Since the sum of $S.m$ in the scheduling queue can be as large as $2K$, a new HOL cell at any queue has to wait, at, most $K(N-1) + 2K$ cell forwarding times.

The ORR arbiter can be used to implement the RR arbitration in the CICQ switch. Each of N CPs, CP_i where $i = 1, 2, ..., N$ associated with VOQ_i sends one bit of CP status, F_i, to its input port. For a CP buffer size of $3K$ cells, an F_i of 0 is sent if the occupancy at CP_i is below K. An F_i of 1 is sent if the occupancy at CP_i is at or above K (at most $2K$ cells destined for CP_i may be queued in the scheduling queue of the ORR arbiter). Thus, F_i controls the operation of the line 5 of the polling algorithm as *if* $((mark{>}0)$ *and* $(F_i == 0))$.

8.6 Future Directions – The CICQ with VCQ

This section outlines one possible solution to reducing CP buffer size due to a large RTT (measured in cell times) between the input ports and crossbar switch fabric in a CICQ switch. Virtual crosspoint queues (VCQs) are proposed as a means of reducing the amount of required memory within the crossbar.

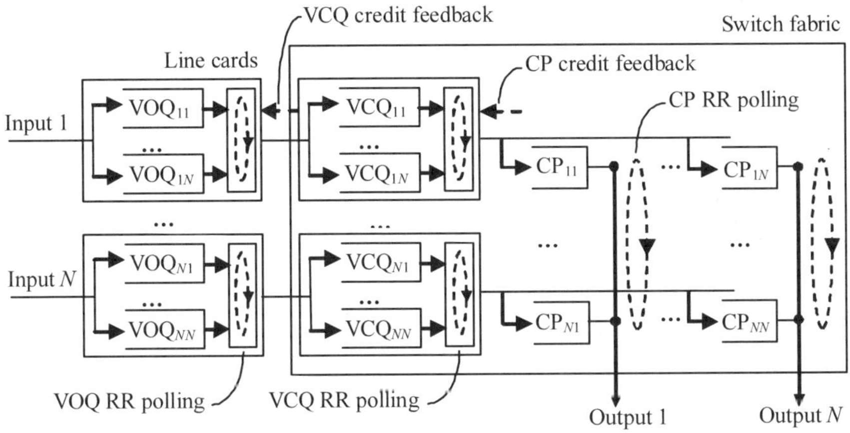

Figure 8.16. CICQ switch with VCQ

8.6.1 Design of Virtual Crosspoint Queueing (VCQ)

The VCQ concept places virtual queues at the edge of (and within) the crossbar switch fabric. Figure 8.16 shows VCQs in a CICQ switch. Similar to an existing CICQ switch, there is buffering and scheduling at the input ports and CP buffering and scheduling internal to the switch fabric. In addition, there are N memory units and N VCQ schedulers internal to the switch fabric physically near to the crossbar. The memory unit is partitioned into N logical queues each dedicated to a unique CP in the form of VCQ. Each memory unit is dynamically shared among N VCQs. Thus, the size of the memory unit is the sum of VCQ_size_{ij} for $i = 1, 2, \dots N$, where VCQ_size_{ij} is a size of VCQ for input i and output j. Two credit-based flow control mechanisms are used as shown in Figure 8.17. The flow control between VCQ and CP buffers uses a unique credit, CP_credit_{ij} per CP buffer, CP_{ij}, while the flow control between an input port i and a set of N VCQs uses a common credit, $VCQs\_credit_i$. Separate arbiters are used to schedule packets buffered in the VOQ, VCQ, and CP, each of which can be an RR poller. For the CICQ switch without VCQ, each buffer size at CP_{ij}, CP_credit_{ij}, must be large enough to hold RTT cells and requires RTT credits to guarantee 100% throughput for completely unbalanced traffic where an input i sends its entire traffic to a single output port j. The CICQ with VCQ can dynamically allocate its VCQs to buffer completely unbalanced traffic. Thus, the CICQ–VCQ switch with CP_size_{ij} of one cell needs to hold only RTT cells in the VCQ buffers to guarantee 100% throughput. The CP_size_{ij} and sum of VCQ_size_{ij} that provide acceptable performance for both balanced and unbalanced traffic are investigated in the next section.

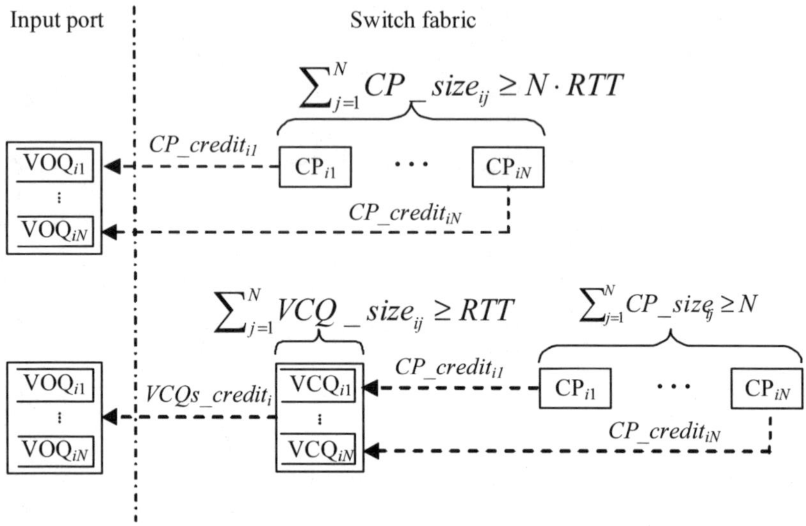

Figure 8.17. CICQ with and without VCQ

8.6.2 Evaluation of CICQ Cell Switch with VCQ

The performance of a 16-port CICQ switch with and without VCQ was evaluated using simulation. For all experiments, the RTT value was set to 64 cell times.

Bernoulli experiment (balanced): Bernoulli arrival of cells (as described in Section 8.3) with uniformly selected outputs. Offered load was ranged from 50% to 98%. The CP buffer size was set to 16 and 64 cells for the CICQ switch with and without VCQ, respectively. The VCQ memory unit size was varied from RTT to twice RTT. This experiment evaluates the switching delay of a switch with a large RTT for smooth traffic.

Bernoulli experiment (unbalanced): Bernoulli arrivals of cells with unbalanced traffic as used by Rojas-Cessa *et al.* [15]. For input i, output j, unbalanced probability ω, and offered input load ρ, the traffic load from input i to output j, $\rho_{i,j}$, was given by

$$\rho_{i,j} = \rho \left(w + \frac{1-\omega}{N} \right)$$

if $i = j$ and otherwise by

$$\rho_{i,j} = \rho \left(\frac{1-\omega}{N} \right).$$

The offered traffic is uniform when $\omega = 0$ and is completely directional from input i to j when $\omega = 1$. A CP buffer size is varied to hold up to RTT cells. This experiment evaluates the throughput of a switch with a large RTT under unbalanced traffic. This traffic model has been used to evaluate the impact of CP size on the performance of the CICQ switch in [25–28, 46, 47].

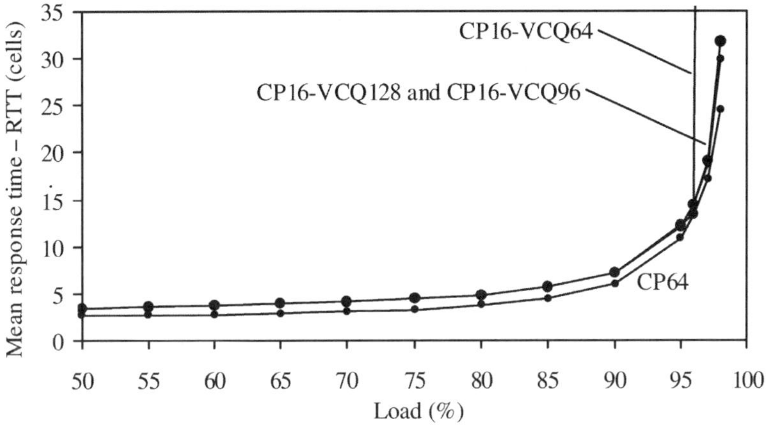

Figure 8.18. Mean response time – RTT (balanced Bernoulli arrivals)

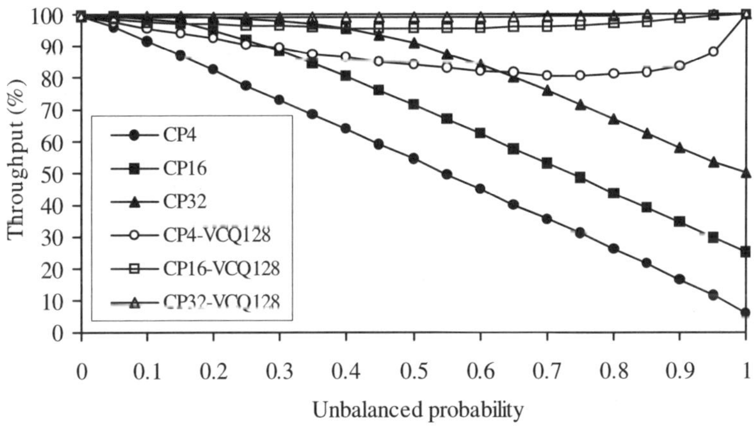

Figure 8.19. Throughput of CICQ switch with and without VCQ

Figure 8.18 shows the result of the balanced Bernoulli experiment in which values of mean delay minus RTT is plotted. The CICQ switch with VCQ buffers of size 64 cells became unstable at 97% offer load. Both of the CICQ switches with VCQs of size 96 and 128 cells were stable for all offered loads and had comparable mean delay to that of the CICQ switch with a CP size of 64 cells. About one cell time difference (due to the extra buffering stage by VCQ) was observed at low offered load. Figure 8.19 shows the result of the unbalanced Bernoulli experiment. All models achieved close to 100% throughput when traffic was uniformly distributed. The throughput of the CICQ–VCQ switch was close to 100% when $\omega = 1$ regardless of

CP size. This was expected because

$$\sum_{j=0}^{N} VCQ\_size_{ij} \geq RTT.$$

Throughput of the CICQ switch with CP size of 4, 16, and 32 cells decreased significantly as traffic became unbalanced, becoming equal to $(CP\_size_{ij}/\text{RTT}) \cdot 100$ when $\omega = 1$, as expected. The throughput of the CICQ–VCQ switch with CP size of 16 and 32 cells was close to 100% throughput for all ω. For VCQ size of 128 cells with CP size of 16 cells, memory savings in each row of the crossbar of (64 cells ·16 CPs) – (16 cells · 16 CPs + 128 cells)=640 cells is achieved. This is a 62.5% reduction in required memory size for the switch fabric, compared to the CICQ switch with CP size of 64 cells.

8.7 Summary

In summary, the CICQ switch is a merging of the buffered crossbar and VOQ switch. The CICQ switch exhibits very good performance and due to its simple schedulers is very likely to be highly scalable. The CICQ switch does exhibit instability for a schedulable, asymmetric traffic load. For any two ports arbitrarily identified as ports 1 and 2, let $\lambda_1 = \lambda_{11} + \lambda_{12}$, $\lambda_{21} = \lambda_{12}$, and $\lambda_{22} = 0$. Within a region of $\lambda_{11} > 0.5$ and high offered traffic load, instability occurs. This instability condition is not limited to a two-port switch, but can occur between any two ports of a large switch. This and other instability conditions exist also in iSLIP VOQ switches. Speedup and other methods can be used to overcome the instability (see, for example, [23]). Clearly, additional work needs to be done to investigate the stability of the CICQ switch. Ideally, better analytical models for CICQ switch performance are needed. Additional work is also needed in reducing the constraints caused by feedback delay from the CP buffers to the input port schedulers. Section 8.6 described one possible direction. Rojas-Cessa *et al.* have described another direction based on load-balancing a switch [47]. It is encouraging to see the continued interest and ongoing work to better understand and further develop the CICQ switch.

References

1. Y. Tamir and G. Frazier (1988) High performance multi-queue buffers for VLSI communications switches. *Proc. of Computer Architecture* 343-354
2. T. Anderson, S. Owicki, J. Saxe, and C. Thacker (1993) High-speed switch scheduling for local-area networks. *ACM Transactions on Computer Systems* 11:319-352
3. N. McKeown (1995) Scheduling algorithms for input-queued switches. PhD Thesis, University of California at Berkley
4. S. Chuang, A. Goel, N. McKeown, and B. Prabhakar (1999) Matching output queueing with a combined input/output-queued switch. *IEEE JSAC* 17:1030-1039.
5. M. Nabeshima (2000) Performance evaluation of a combined input- and crosspoint-queued switch. *IEICE Transactions on Communications* E83-B:737-741

6. R. Bakka and M. Dieudonne (1982) Switching circuit for digital packet switching network. *United States Patent 4,314,367*
7. S. Nojima, E. Tsutio, H. Fukuda, and M. Hashimoto (1987) Integrated services packet network using bus matrix switch. *IEEE JSAC* 5:1284-1292
8. A. Gupta, L. Barbosa, and N. Georganas (1991) 16x16 limited intermediate buffer switch module for ATM networks. *Proc. of IEEE GLOBECOM* 939-943
9. A. Gupta, L. Barbosa, and N. Georganas (1992) Limited intermediate buffer switch modules and their interconnection networks for B-ISDN. *Proc. of IEEE ICC* 1646-1650
10. Y. Doi and N. Yamanaka (1993) A high-speed ATM switch with input and cross-point buffers. *IEICE Transactions on Communications*, E76-B:310-314
11. E. Re and R. Fantacci (1993) Performance evaluation of input and output queueing techniques in ATM switching systems. *IEEE Transactions on Communications* 40:1565-1575
12. D. Stephens and H. Zhang (1998) Implementing distributed packet fair queueing in a scalable switch architecture. *Proc. of IEEE INFOCOM* 282-290
13. V. Singhal and R. Le (2000) High-speed buffered crossbar switch design using virtex-EM devices (http://www.xilinx.com/xapp/xapp240.pdf)
14. K. Yoshigoe and K. Christensen (2001) A parallel-polled virtual output queued switch with a buffered crossbar. *Proc. of IEEE Workshop on High Performance Performance Switching and Routing* 271-275
15. R. Rojas-Cessa, E. Oki, Z. Jing, and H. Chao (2001) CIXB-1: combined input-one-cell cross point buffered switch. *Proc. of IEEE Workshop on High Performance Switching and Routing* 324-329
16. R. Rojas-Cessa, E. Oki, and H. Chao (2001) CIXOB-k: combined input-cross point-output buffered packet switch. *Proc. of IEEE GLOBECOM* 2654-2660
17. J. Chao (2000) Saturn: a terabit packet switch using dual round-robin, *IEEE Communication Magazine* 38(12):78-84
18. M. Han, D. Kwak, and B. Kim (2003) Desynchronized input buffered switch with buffered crossbar. *IEICE Transactions on Communications* E86-B:2216-2219
19. B. Yoon, M. Han, H. Lee, B. Kim, and W. Kim (2004) Exhaustive output arbitration of input buffered switch with buffered crossbar. *ETRI Journal* 26:505-508
20. I. Radusinovic and M. Pejanovic (2002) Impact of scheduling algorithms on performances of buffered crossbar switch fabrics. *Proc. of IEEE ICC* 2416-2420
21. T. Javadi, R. Magill, and T. Hrabik (2001) A High-throughput scheduling algorithm for a buffered crossbar switch fabric. *Proc. of IEEE ICC* 1581-1591.
22. L. Mhamdi and M. Hamdi (2003) Practical scheduling algorithms for high-performance packet switches. *Proc. of IEEE ICC* 1659-1663
23. K. Yoshigoe and K. Christensen (2003) An evolution to crossbar switches with virtual output queueing and buffered cross points. *IEEE Network* 17(5):48-56
24. N. Gunther, K. Christensen, and K. Yoshigoe (2003) Characterization of the burst stabilization protocol for the RR/RR CICQ switch. *Proc. of IEEE LCN* 260-269
25. R. Rojas-Cessa and E. Oki (2003) Round-robin selection adaptable-size frame in a combined input-cross point buffered switch. *IEEE Communications Letters* 7:555-557
26. R. Rojas-Cessa (2003) Round-robin with adaptable-size frame arbitration for input-cross point buffered switch. *IEEE Communications Society* 7:1113-1117
27. R. Rojas-Cessa (2004) High-performance round-robin arbitration schemes for input-cross point buffered switch. *Proc. of IEEE Workshop on High Performance Switching and Routing* 167-171
28. L. Mhamdi and M. Hamdi (2003) MCBF: a high-performance scheduling algorithm for buffered crossbar switches. *IEEE Communications Letters* 7:451-453

29. X. Zhang and L. Bhuyan (2004) An efficient scheduling algorithm for combined input-cross point-queued (CICQ) switches. *Proc. of IEEE GLOBECOM* 1168-1173
30. M. Lin and N. McKeown (2005) The throughput of a buffered crossbar switch. *IEEE Communications Letters* 9:465-467
31. S. Motoyama, M. Arantes (2002) IP switch with distributed scheduling. *Electronics Letters* 38:392-393
32. Q. Duan and J. Daigle (2002) Resource allocation for quality of service provision in buffered crossbar switches. *Proc. of IEEE ICCCN* 509-513
33. N. Chrysos and M. Katevenis (2003) Weighted fairness in buffered crossbar scheduling. *Proc. of IEEE HPSR* 17-22
34. R. Magill, C. Rohrs, and R. Stevenson (2003) Output queued switch emulation by fabrics with limited memory. *IEEE JSAC* 21:606-615
35. L. Mhamdi and M. Hamdi (2003) Output queued switch emulation by a one-cell-internally buffered crossbar switch. *Proc. of IEEE GLOBECOM* 3688-3693
36. S. Chuang, S. Iyer, and N. McKeown (2005) Practical algorithms for performance guarantees in buffered crossbars. *Proc. of IEEE INFOCOM* 981-991
37. P. Giaccone, E. Leonardi, and D. Shah (2005) On the maximal throughput of networks with finite buffers and its application to buffered crossbars. *Proc. of IEEE INFOCOM* 971-980
38. S. He, S. Sun, W. Zhao, Y. Zheng, and W. Gao (2005) Smooth switching problem in buffered crossbar switches. *Proc. of SIGMETRCS* 386-387
39. K. Yoshigoe, K. Christensen, and A. Roginsky (2005) Performance evaluation of new scheduling methods for the RR/RR CICQ switch. *Computer Communications* 28:417-428
40. M. Katevenis and G. Passas (2005) Variable-size multipacket segments in buffered crossbar (CICQ) architectures. *Proc. of IEEE ICC* 999-1004
41. F. Abel, C. Minkenberg, R. Luijten, M. Gusat, and I. Hiadis (2002) A four-terabit single-stage packet switch with large round-trip time support. *IBM Research Report RZ 3430 (#93609)*
42. R. Luijten, C. Minkenberg, and M. Gusat (2003) Reducing memory size in buffered crossbars with large internal flow control latency. *Proc. of IEEE GLOBECOM* 3683-3687
43. N. Chrysos (2003) Design issues of variable-packet-size, multiple-priority buffered crossbars. *Technical Report FORTH-ICS/TR*-325
44. N. Chrysos and M. Katevenis (2004) Multiple priorities in a two-lane buffered crossbar. *Proc. of IEEE GLOBECOM* 1180-1186
45. F. Gramsamer, M. Gusat, and R. Luijten (2003) Flow control scheduling. *Microprocessors and Microsystems* 27:233-241
46. M. Katevenis, G. Passas, D. Simos, I. Papaefstathiou, and N. Chrysos (2004) Variable packet size buffered crossbar (CICQ) switches. *Proc. of IEEE ICC* 1090-1096
47. R. Rojas-Cessa, Z. Dng, and Z. Guo (2005) Load-balanced combined input cross point buffered packet switch and long round-trip times. *IEEE Communications Letters* 9:661-663
48. H. Schwetman (1996) CSIM18 - the simulation engine. *Proc. of the 1996 Winter Simulation Conference* 517-521
49. M. Goudreau, S. Kolliopoulos, and S. Rao (2000) Scheduling algorithms for input-queued switches: randomized techniques and experimental evaluation. *Proc. of IEEE INFOCOM* 1634-1643
50. N. McKeown (1999) The iSLIP scheduling algorithm for input-queued switches. *IEEE/ACM Transactions on Networking* 7:188-201

51. N. McKeown and T. Anderson (1998) A quantitative comparison of iterative scheduling algorithms for input-queued switches. *Computer Networks and ISDN Systems* 30:2309-2326
52. K. Christensen (2005) *Home page for Ken Christensen* (http://www.csee.usf.edu/~christen)
53. A. Mekkittikul and N. McKeown (1996) A starvation-free algorithm for achieving 100% throughput in an input-queued switch. *Proc. of IEEE ICCCN* 226-231
54. G. Nong, M. Hamdi, and K. Letaief (1999) Efficient scheduling of variable-length IP packets on high-speed switches. *Proc. of IEEE GLOBECOM* 1407-1411
55. M. Marsan, A. Bianco, P. Giaccone, E. Leonardi, and F. Neri (2002) Packet-mode scheduling in input-queued cell-based switches. *IEEE/ACM Transactions on Networking* 10:666-678
56. J. Delgado-Frias and J. Nyathi (1998) A VLSI high-performance encoder with priority lookahead. *Proc. of the 8th Great Lakes Symposium* on VLSI 59-64
57. P. Gupta and N. McKeown (1999) Design and implementation of a fast crossbar scheduler. *IEEE Micro* 19:pp. 20-28
58. J. Chao, C. Lam, and X. Guo (1999) A fast arbitration scheme for terabit packet switches. *Proc. of IEEE GLOBECOM* 1236-1243
59. E. Shin, V. Mooney III, and G. Riley (2002) Round-robin arbiter design and generation. *Proc. of the 15th International Symposium on System Synthesis* 243-248
60. S. Zheng, M. Yang, J. Blanton, P. Golla, and D. Verchere (2002) A simple and fast parallel round-robin arbiter for high-speed switch control and scheduling. *Proc. of the 45th Midwest Symposium on Circuits and Systems* 671-674
61. S. Huajin, G. Deyuan, Z. Shengbing, and W. Danghui (2002) Design fast round robin scheduler in FPGA. *Proc. of IEEE Communications, Circuits and Systems and West Sino Expositions* 1257-1261
62. A. Motoki, S. Kamiya, R. Ikematsu, and H. Ozaki (2001) Group-pipeline scheduler for input-buffer switch. *Proc. of IEEE International Conference on ATM and High Speed Intelligent Internet Symposium* 158-162
63. A. Smiljanic, R. Fan, and G. Ramamurthy (1999) RRGS-round-robin greedy scheduling for electronic/optical terabit switches. *Proc. of IEEE GLOBECOM* 1244-1250
64. E. Oki, R. Rojas-Cessa, and J. Chao (2002) PCRRD: a pipeline based concurrent round-robin dispatching scheme for clos network switches. *Proc. of IEEE ICC* 2121-2125
65. K. Yoshigoe, K. Christensen, and A. Jacob (2003) The RR/RR CICQ switch: hardware design for 10-Gbps link data rate. *Proc. of IEEE IPCCC* 481-485
66. Xilinx Inc. (2005) Xilinx WebPACK 4.2 (http://www.xilinx.com/xlnx/xil_prodcat_landingpage.jsp?title=ISE+WebPack)
67. K. Yoshigoe, L. Christensen, and A. Roginsky (2003) Design of a high-speed overlapped round robin (ORR) arbiter. *Proc. of the IEEE LCN* 638-639

9

Time–Space Label Switching Protocol (TSL-SP)

Anpeng Huang, Biswanath Mukherjee, Linzhen Xie, and Zhengbin Li

Department of Computer Science, The University of California at Davis.
{hapku,mukherje,tydxlz,lizhb}@cs.ucdavis.edu

9.1 Introduction

As the volume of data traffic continues to increase, how to reasonably utilize the offered bandwidth in an optical wavelength-division-multiplexing (WDM) network is becoming a major challenge. An efficient switching mechanism may lend itself to realize this goal. Basically, there are three switching schemes for WDM networks: circuit switching, packet switching, and burst switching. Optical circuit switching (OCS) is generally implemented as wavelength routing, which is a popular scheme in WDM networks today, but its usage of bandwidth may not be efficient for highly bursty traffic. A longer-term strategy for the network evolution is to employ optical packet switching (OPS), which provides better resource utilization, higher functionality, and finer switching granularity. One of the major challenges of OPS is that there is no optical equivalent of a random access memory (RAM). Optical burst switching (OBS) tries to combine the merits of optical circuit switching and optical packet switching while avoiding their shortcomings [1–7].

Optical burst switching (OBS) allows switching of data channels entirely in the optical domain by performing resource allocation in the electronic domain, as shown in Figure 9.1. In an OBS network, a control packet is sent into the network by the source node prior to the corresponding data burst. The control packet and the corresponding data burst are launched at the source node separately with an offset time. The control packet (also called a burst control packet (BCP)) contains the necessary information for routing the data burst through the optical core network, and it is sent in an out-of-band control channel (possibly a separate wavelength or a subcarrier). The control packet is processed electronically at each intermediate node (which also contains the switching fabric which is an optical cross-connect) to make routing decisions (outgoing interface and wavelength), and the optical cross-connects are configured to switch the data burst entirely in the optical domain, thereby removing the electronic bottleneck in the end-to-end routing path.

In order to efficiently utilize the bandwidth, to reduce the latency, and to provide the data-transmission transparency, some OBS signaling protocols have been proposed [8–11]. But these protocols are all similar in that the signaling function is

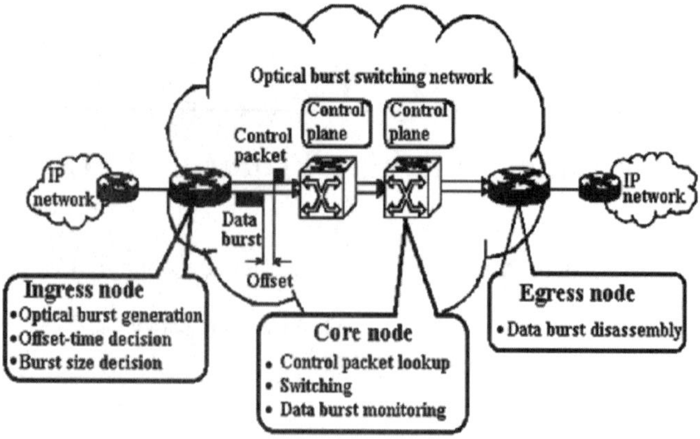

Figure 9.1. Optical burst switching (OBS) paradigm

separated from the routing function and therefore introduces constraints on the coordination among the routing, protection, and scheduling of bursts in an OBS network. In optical networks, routing and signaling could be considered together instead of separated control components [1].

In this chapter, a novel Time–space Label Switching Protocol (TSL-SP) is proposed for Optical Burst Switching Networks. The concepts of "time label" and "space label" are defined first as two dimensions to the switching problem. In order to illustrate the operational principles of the two-dimensional label-switched paths in TSL-SP, a new approach of orthogonal time-space coordinates is introduced, in which the vertical coordinate is the space label and the horizontal coordinate is the time label. In the orthogonal time-space coordinates, network resources can be efficiently assigned by the TSL-SP, and therefore the routing and signaling functions can be well integrated.

This chapter is organized as follows. In Sections 9.2 and 9.3, the concepts of *time label* and *space label* are defined. In Section 9.4, TS-LSP is illustrated. In Section 9.5, the performance of TSL-SP vs. conventional signal protocols is compared. Finally, the chapter is summarized in Section 9.6.

9.2 Time Label

Basically, the time-label approach is a resource-reservation mechanism for optical networks, and it is one of signaling protocols of OBS. In conventional signaling protocols of OBS (see Figure 9.2) [9], the resource reservation is implemented by the offset time between the control packet and the data burst. At the source node, the offset time is not smaller than the sum of the expected control delays in each node along the routing path, *i.e.* $T \geq \sum_{h=1}^{n} \delta(h)$, where $\delta(h)$ is the expected control delay (*e.g.* the processing time incurred by the control packet at node h, and n is the

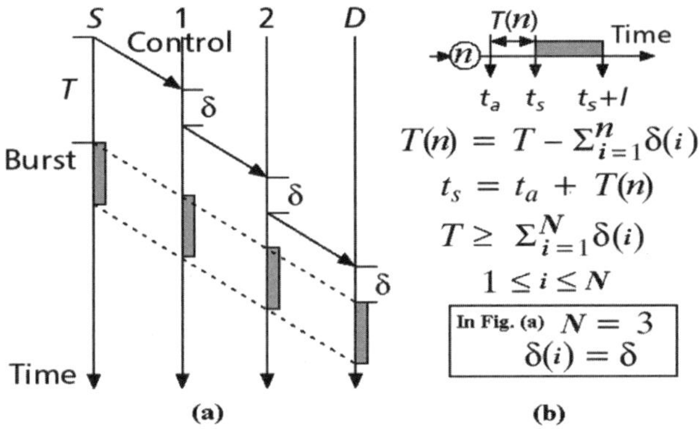

Figure 9.2. Time label

total number of traversed nodes) [9]. In current generation of WDM switches, the signaling-protocol processing time is a few microseconds, and the cross-connect cut-through time is a few milliseconds [8]. There could be a certain amount of deviation between the estimated and actual control-delay time, which is on the order of a few microseconds. This deviation may result in a data-burst loss or wastage of the link resource.

For dealing with the above problem, we propose the new concept of *time label.* The fundamental principle of time-label switching is that, if the source node of a burst (as well as all other nodes) has global information on the time-wise occupancy status of all the links in the network, then the source node could pick an appropriate time such that the entire data burst can traverse the path through intermediate nodes to the destination without being dropped at an intermediate node due to contention. Thus, instead of launching a burst right away into the network as in standard OBS, the launching of the burst into the network may be delayed until an opportunistic moment to ensure smooth passage of the burst to the destination. But a contention may still occur because of slightly out-of-date information on link status, *e.g.* when two nodes launch bursts into the network nearly simultaneously while using a common link, which was expected to be available at the appropriate time by both the nodes.

The time-label mechanism can be implemented as follows. The instant when a data burst arrives at an intermediate node is called the time label. As shown in Figure 9.2(a), the control packet is sent by the source node at time T_c, and the corresponding data burst is sent at time T_b. H_i represents the propagation delay that a data burst will encounter on the ith link (or hop), where i goes from 1 to n on this n-hop path. At the source node, the time label is referred to as T_b. At the first intermediate node, the time label is $T_1 = T_b + H_1$. The reserve-fixed-duration (RFD) at this node is from T_1 to $T_1 + L$, where L is the length of the corresponding data burst. At the second intermediate node, the time label is $T_2 = T_b + H_1 + H_2$, and RFD is from T_2 to

$T_2 + L$. At the nth node, the time label is $T_n = T_b + \sum_{i=1}^{n} H_i$, and RFD is from T_n to $T_n + L$, and so on.

At the source node, $\triangle T$ must be not smaller than the offset time T of conventional OBS signaling protocols, *i.e.* $\triangle T \geq T$, where $\triangle T = T_b - T_c$. Because the control packet provisions network resources for the corresponding data bursts, the control packet must be processed before the corresponding data burst arrives at each node. Because no optical buffer is applied in such an OBS network, the data burst will be transparently propagated to the destination at light speed without extra delay. Thus, as long as the time label T_b at the source is determined, the corresponding time label of each intermediate node will be exactly determined. Therefore, T_b, T_1, T_2, ..., T_n are taken as the time labels. Each time label is a local time identifier (ID) at its own node, and it is also a remote time ID at the other nodes located on the routing path. An ordinal label stacking consists of the time labels of the routing path, which is shown in Figure 9.2(b). Along the routing path, the corresponding time label of each node is turning from a remote time ID into a local time ID, which is on the top of the time-label stack. A controller in an intermediate node reads the local time ID.

Why should one replace the offset time by time label for signaling in an OBS network? In optical networks, optical propagation time per kilometer is about 5 μs. In current optical networks, the deviation of optical propagating time in each hop (or fiber link) is in the order of 10^{-8}s, which is much smaller than the deviation magnitude caused by the control delay of conventional OBS signaling protocols. Another important advantage of the time label is that the corresponding data burst will be triggered at the explicit time label T_b at the source node, which avoids the prediction mechanism of the start and the end of the data bursts in conventional signaling protocols of OBS [8, 9]. Thus, the time label will be an exact mechanism to perform and refine reservation of network resources.

9.3 Space Label

The time label can perform provision of services in the time domain, but how to realize provision in the space domain is equally important. Specifically, given that there are multiple paths available from the source to the destination in a mesh network, which path should be chosen for quickest delivery of the data burst to the destination?

Shown in Figure 9.3 are routing-adjacency relationships between three neighboring nodes. In the routing protocol, each node and its ports can be distinguished from one another by some specific identifiers. Those corresponding identifiers are called space labels. In Figure 9.3, the $(N + 1)$th node is represented by Router ID: $N + 1$, and each intermediate node of the routing path has an input port and an output port, such as $(N + 1)$th node with an input port ID $P2$ and an output port ID $P3$. The labels $P2$ and $P3$ are local for the $(N + 1)$th node, but they are also the remote labels for the Nth node. In Figure 9.3, the arrowheads show the traffic direction. In the optical network, a fiber link (or hop) can be uniquely identified by two port IDs. For example, in Figure 9.3, a link $L1$ is uniquely identified by $P1$ and $P2$. Each

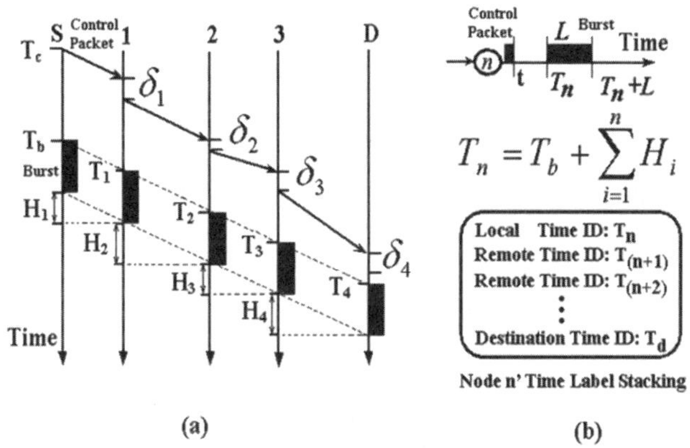

Figure 9.3. Space label

port ID contains three necessary pieces of information: (1) switching information, (2) maximum (or reservable) bandwidth, and (3) data encoding information.

Also optional information may be attached in the descriptor. For example, the allowable maximum burst length could be added. An example of a port ID is as follows: Switching Type = OBS-1, Encoding = Ethernet 802.3, Max Bandwidth [0] = 1.0 Gbps for priority 0, Max Burst Length =300 ms.

By space labels, a port ID can be used to identify further fine-grain information within each node. All space labels of a routing path form a label stacking, which is shown in Figure 9.3. The space-label stacking operates in the same way as time-label stacking. In each node, the controller reads the space ID along with time ID, and the resource reservation is implemented in the nodes. Along the routing path, the local space ID is located at the top of the stack, and it is read and then stripped off. At the next node, the first remote space ID appears at the top of the stack and becomes a local space ID.

Space label can implement provisions for services in the space domain, and is part of the routing information. Note that each wavelength channel could be labeled based on the switching port identifiers in a WDM network.

9.4 Time–Space Label Switching Protocol (TSL-SP)

The time–space label (TSL) consists of time and space labels, and hence it is two-dimensional. The time label avoids the impact of control delay for provisioning of services. But node synchronization is needed in the time-label scheme, as in the JET procedure [9], and it can be more easily implemented than bit or packet synchronization. The TSL carries the routing and signaling information. Also the topology and link-status information of the network should be collected for efficient utiliza-

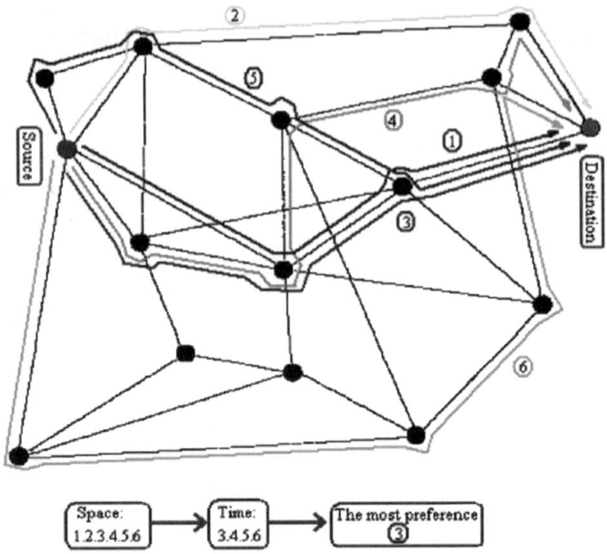

Figure 9.4. An example of the time–space routing algorithm

tion of network resources. Most importantly, a time–space label is network-wide and globally unique.

The Time–Space Routing (TSR) algorithm's objective is to find a shortest-path tree in which the necessary network resources are not only free in the space dimension but also available in the time dimension. The algorithm consists of three major steps:

1. The link-state database of the network is built. Here, the Shortest Path First (SPF) [12] algorithm is adopted to search for such some routing paths between each pair of nodes. These routing paths between node pairs constitute a set in which they are sorted from the shortest path to the longest.
2. Some of the paths are filtered to form a subset if they are available in the time dimension. In this subset, the shortest path is the most preferred, and selected as the primary working path, while other paths in the subset could be taken as candidates for backup paths. Based on the primary working path, the SPF tree can be formed, and it can serve as a map for data-burst routing.
3. At the source node, a routing table is built based on the corresponding SPF tree, which consists of space and time labels.

In the algorithm, the selected shortest routing path may not be the shortest routing path in the space dimension, but it must be available in the time dimension. Here, the meaning of "shortest" is the shortest path among all paths that are available in the time dimension. The rule guarantees that the network resources assigned are efficient and effective in time and space dimensions. TSR is illustrated in Figure 9.4, in which the set of routing paths are searched and sorted according to the lengths in hops (and ties between some equal-hop paths are broken randomly). In Figure 9.4, six routing paths between the source and destination nodes are considered. But, in the time dimension, it may be the case that there are only four routing paths (3,4,5,6) currently available among the set of routing paths. In the subset of four routing paths, path number 3 is the most preferred because it is the best available shortest path in both time and space dimensions.

The fundamental goal of TSL-SP is to coordinate the signaling and routing functions organically. In TSL-SP, the routing function uses a label-based routing protocol to distribute and maintain information about the topology and resources of the network. The routing protocol is the means by which non-local resource information is distributed. The topology and resources of the network will be taken into account as parameters for the path algorithm to calculate a label-switched path (LSP). In TSL-SP, the signaling function is the procedure through which service provisioning occurs. The service provisioning includes LSP establishment, LSP deletion, and LSP modification. In TSL-SP, after comparing the time-label information of the routing table at each intermediate node with the time information in the link-state database, the routing path is confirmed if there will be no contentions. If such a routing path is not available, TSL-SP searches for the second preferred path and so on, until such a routing path is finally confirmed. Then, TSL-SP distributes the time–space label (TSL) along the confirmed routing path, and the network resources are assigned accordingly. Thus, a LSP in time and space dimensions can be set up by TSL-SP, which is called the time–space label switched path (TS-LSP).

In order to demonstrate the principles of the TSL-SP, orthogonal time–space coordinates in which the vertical coordinate is the space label and the horizontal coordinate is the time label are shown in Figure 9.5.

In Figure 9.5, there are some assumptions: a mesh network with 10 nodes, four ports per node with two inputs and two outputs, and the nodes in the network are connected to each other by a single fiber link. In the orthogonal time–space coordinates, a crossing point between time label and space label is an intermediate node. Because the propagating time of a burst through the intermediate node is neglected in TS-LSP, the input port ID and output port ID in an intermediate node correspond to the same time ID in Figure 9.5. In the intermediate node, cut-through switching is accomplished just prior to the arrival of the corresponding data burst. The corresponding data burst is forwarded and transmitted along the TS-LSP. In Figure 9.5, each TS-LSP can be determined by the corresponding time and space labels.

In particular, TS-LSP dramatically reduces the blocking probabilities of routing and forwarding. Resource provisioning of conventional signaling protocols is carried out on a single dimension, *i.e.* either the time or space dimension. However, resource provisioning in only one dimension could lead to a conflict of resource provisioning in the other dimension. The probabilities of routing failure and burst loss are higher under heavy traffic if only one-dimensional resource provisioning is adopted. In Figure 9.5, the hop of $H(5-7)$ is between nodes 5 and 7, and the hop of $H(9-6)$ is between nodes 9 and 6. The hops of $H(5-7)$ in $TS-LSP1$ and $TS-LSP2$ are in the same wavelength channel and the same fiber link. Although the Port IDs

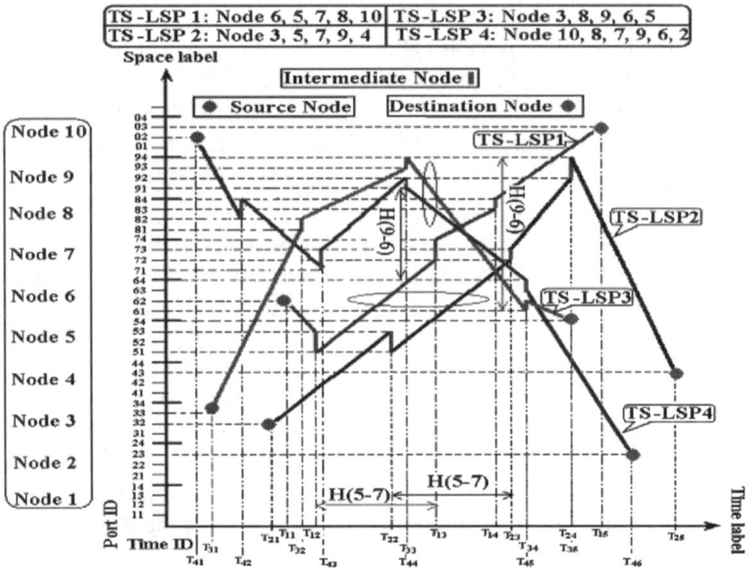

Figure 9.5. Time–space label switching protocol

are the same, the Time IDs are different at nodes 5 and 7. For the hops of $H(9-6)$ in $TS-LSP3$ and $TS-LSP4$, although the Time IDs are the same at nodes 9 and 6, the Port IDs are different at nodes 9 and 6. Thus, the network resources can be statistically utilized in time and space dimensions in TSL-SP. Compared with conventional OBS routing paths, TS-LSP can provide a connection mode for connectionless networks. The connection-oriented nature of TS-LSP is more advantageous when routing paths go through a greater number of hops.

A data burst is carried by a dedicated wavelength in OBS. Thus, a wavelength could be viewed as an implicit label. TSL-SP directs the corresponding data bursts in a dedicated wavelength. TSL-SP could be extended into three dimensions also if the wavelength dimension could also be used as an explicit label in the switching procedure (as shown in Figure 9.6). The three-dimensional label-switching protocol could be further extended into a multi-dimensional switching protocol, *e.g.* finer switching granularities could be additional dimensions.

9.5 Illustrative Results

The features of TSL-SP are described in Figure 9.7. Its characteristics are: one-way reservation, switching cut-through, variable-length payload, out-of-band control, and medium/large switching granularity. The one-way reservation obviously reduces control and procession overhead. The switching cut-through avoids the processing of optical payload. Bandwidth utilization can be improved because of medium/large

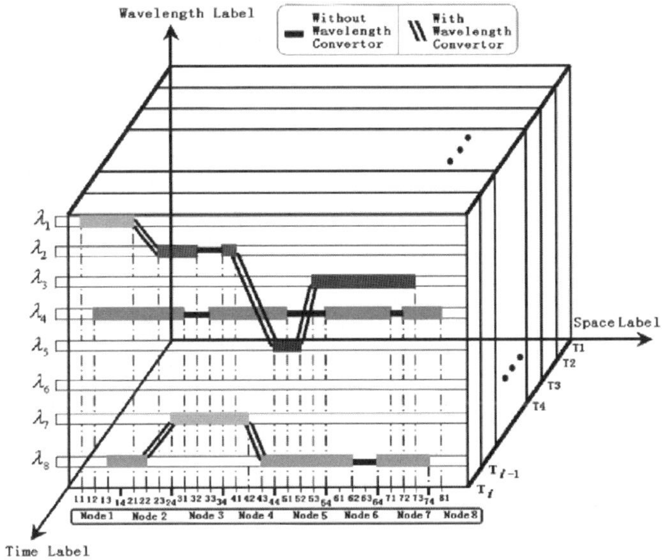

Figure 9.6. Three-dimensional label switching protocol

granularity of bursts. Out-of-band control channel can avoid optical buffer, which is still not very mature today. Finally, the variable length of bursts can provide flexibility.

Although TSL-SP has been proposed for OBS, its modified version can be applied to wavelength routing and packet switching. In Figure 9.7, the bold ellipse shows its applicable fields in optical switching, and the features of all current protocols for optical switching are given in this figure as well.

We evaluate the performance of the proposed TSL-SP on a typical optical backbone network. The network has a randomly generated topology, with 15 nodes and 32 fiber links (see Figure 9.8). There are 32 wavelength channels per fiber link. Self-similar traffic is taken as a resource in which the distributions of the burst inter-arrival time and the burst length follow the Pareto distribution and the exponential distribution, respectively [7]. TSL-SP is compared with conventional OBS signaling protocols, *i.e.* JET and JIT[1] as follows. (Note, SPF is adapted into JET and JIT in our simulations.)

In Figure 9.9, we can see that the switching performance of JIT is better than that of JET. The reason is that JIT has connection acknowledgement and release mechanisms that can guarantee quality of service (QoS) [8], but JET provides best-effort service [9]. When comparing TSL-SP with JET and JIT, we find that the switching

[1] JIT = Just In Time, in which network resources can be reserved for a burst immediately after the arrival of the corresponding control packet; if network resources cannot be reserved at that time, then the control packet is rejected and the corresponding burst is dropped, which is also a typical signal protocol in OBS networks [8, 11]

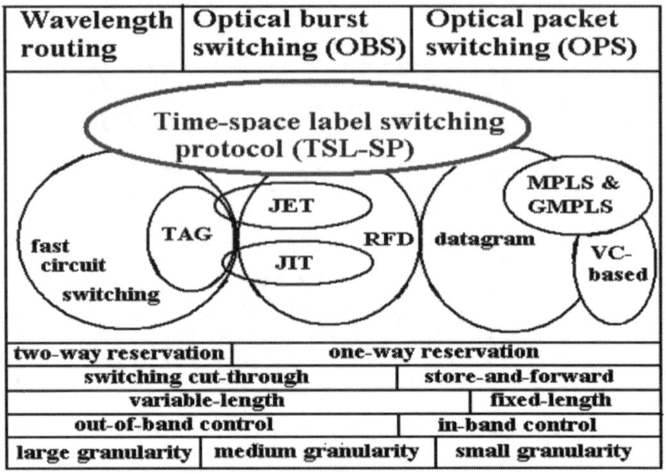

Figure 9.7. Comparison among the switching protocols: TAG - Tell And Go; JET - Just Enough Time; JIT - Just In Time; RFD - Reserve-Fixed Duration; MPLS - Multi-Protocol Label Switching; VC - Virtual Channel

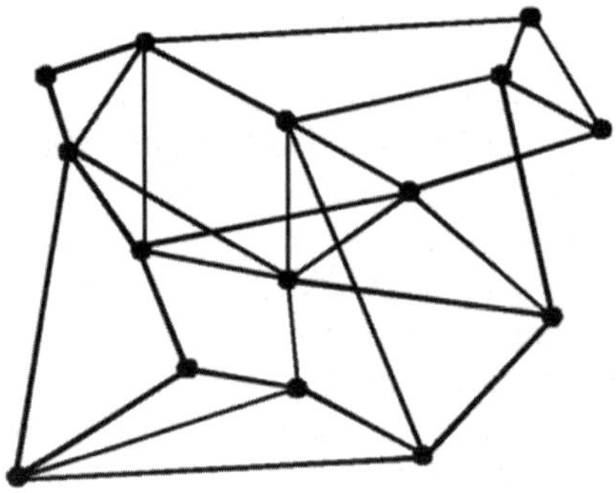

Figure 9.8. A random network topology

performance for TSL-SP is significantly improved. The burst blocking probability of TSL-SP can be reduced by two orders of magnitude compared with JET or JIT in heavy traffic. This is beneficial for statistical utilization of network resources by exploiting the information in the space and the time domains.

Figure 9.10 shows the improvements of link utilization with TSL-SP vs. JIT and JET. As traffic load is increased, greater improvement of link utilization is achieved. Because TSL-SP can provide connection-oriented service for connectionless networks, the network resources can be effectively assigned. Under heavy traffic, the improvement of TSL-SP vs. JET is more than that of TSL-SP vs. JIT because blocking and retransmission probabilities of JET under heavy traffic are higher due to its feature of best-effort service.

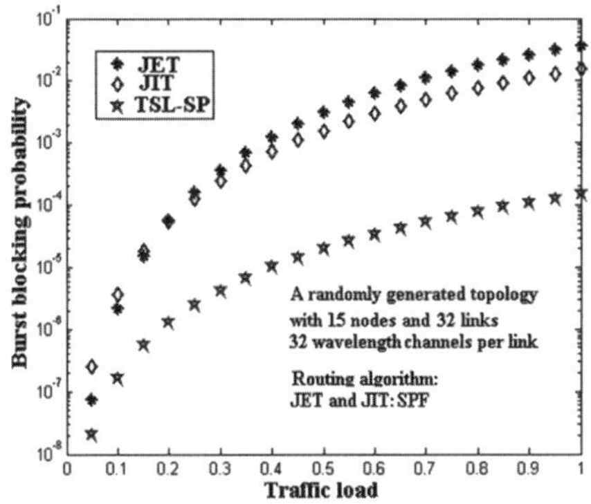

Figure 9.9. Comparison of burst blocking probabilities for different switching protocols

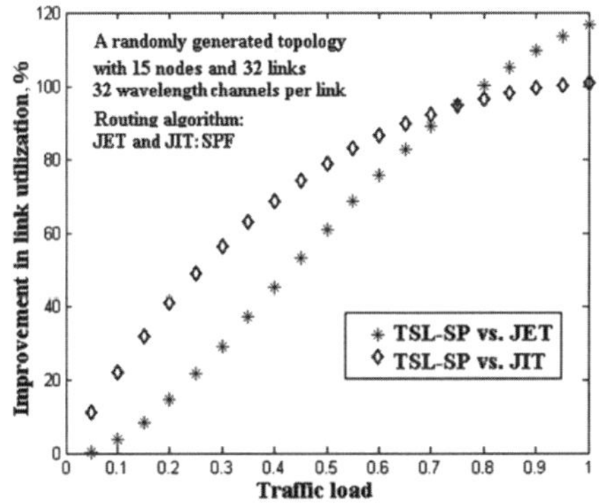

Figure 9.10. Bandwidth utilization improvement of TSL-SP vs. conventional OBS signaling protocols

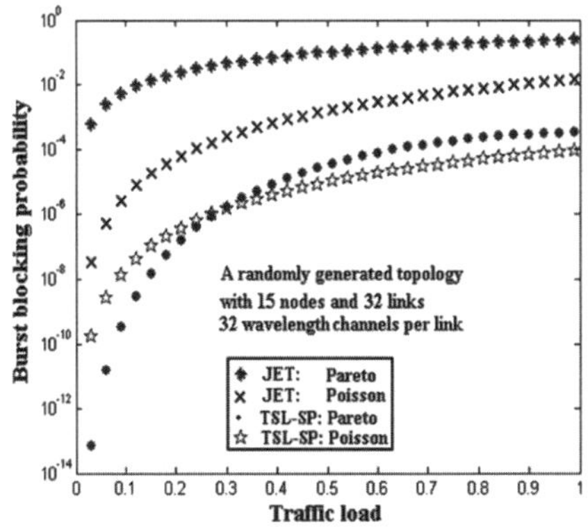

Figure 9.11. Comparison between JET and TSL-SP

Further comparison between JET and TSL-SP is shown in Figure 9.11, where self-similar degree of self-similar traffic (Pareto traffic) is equal to 0.8 ($H = 0.8$). The blocking probability of JET under self-similar traffic is higher (up to 1-3 orders of magnitude) than that of JET under Poisson traffic. This means that JET is not suitable for self-similar traffic. For TSL-SP, the blocking probability of Pareto is lower than that of Poisson under light load, because the heavy tail of self-similar traffic has little effect on network performance under light load. As the volume of traffic increases, the bursty nature is more serious under self-similar traffic, which leads to deterioration of network performance. Thus, the blocking probability of Pareto is higher than that of Poisson under heavy load. But there is no big difference (within one order of magnitude) between Pareto and Poisson for TSL-SP. Thus, TSL-SP helps to mitigate the deterioration of network performance caused by self-similar traffic. Furthermore, with either self-similar traffic or Poisson, the blocking probability of TSL-SP is lower than that of JET.

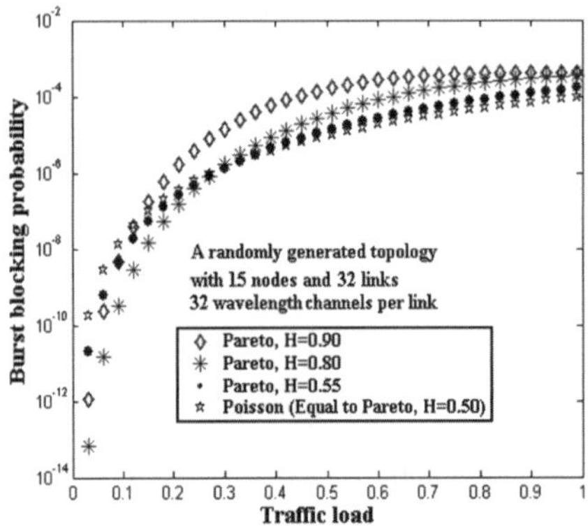

Figure 9.12. Results of TSL-SP under different self-similar degrees of traffic

Figure 9.12 shows the results of TSL-SP under different self-similar degrees of traffic. As the self-similar degree decreases, the blocking probability of TSL-SP goes down, and gradually gets close to that of Poisson traffic. The blocking probabilities for self-similar degrees $H = 0.55$ and $H = 0.5$ are very close. This is because the traffic becomes close to Poisson when H is equal to 0.5. In Figure 9.11, there is a considerable difference between $H = 0.9$ and $H = 0.8$ under moderate and light traffic. But the difference between them reduces when the traffic load is more than 0.8. The main reason behind the phenomenon is that the bottleneck in network resources is the primary factor rather than the self-similar nature that causes the connection to block under heavy traffic load (more than 0.8) and high self-similar degree (more than 0.8).

These illustrative results demonstrate that TSL-SP is an efficient and effective method under self-similar traffic, which is quite different from today's signaling and routing protocols that still suffer from the effects of self-similar traffic.

9.6 Summary

In this chapter, we first defined the time label and space label for transparent optical burst switching networks without extra delay. The fundamental goal of TSL-SP is to bind the signaling with routing functions close together. TSL-SP is a new technology, in which the effectiveness of routing can be achieved by employing the label mechanism. Simulation results show that, when TSL-SP is applied to typical networks, the switching performance can be improved by two orders of magnitude compared with conventional OBS signaling protocols in heavy traffic. Furthermore, a new approach of orthogonal time–space coordinates is proposed to analyze the assignment of networks resources. Based on its two-dimensional feature, the network resources can be utilized more efficiently. In particular, it is effectively immunized from the effect of self-similar traffic.

As the traffic nature changes from being voice-dominant to data-dominant, the self-similar nature of network traffic becomes bursty at all time scales. Thus, under reasonable switching protocols and mechanisms, burst switching could be suitable for today's bursty traffic where the duration of each burst is neither long enough to be suited to circuit switching nor short enough to fit into a single packet. Self-similar cluster switching is a related topic discussed elsewhere [14].

Therefore, TSL-SP is one of the most promising protocols for Optical Burst Switching Networks. For more information on this topic, the interested reader is referred to [13–15].

References

1. K. H. Liu, *IP over WDM*, John Wiley & Sons, New York, pp. 147-151, 2002.
2. L. Xu, H. G. Perros, and G. Rouskas, Techniques for optical packet switching and optical burst switching, *IEEE Commun. Mag.*, **39**, no. 1, pp. 136–142, Jan. 2001.
3. S. Verma, H. Chaskar, and R. Ravikanth, Optical burst switching: a viable solution for terabit IP backbone, *IEEE Network*, **14**, no. 6, pp. 48–53, Nov. 2000.
4. J. White and M. Zukerman, A framework for optical burst switching network design, *IEEE Commun. Letters*, **6**, no. 6, pp. 268–270, June 2002.
5. A. Mokhtar and M. Azizoglu, Adaptive wavelength routing in all-optical networks, *IEEE/ACM Transactions on Networking*, **6**, no. 2, pp. 197–206, Apr. 1998.
6. M. Yang, S. Q. Zheng, and D. Verchere, A QoS supporting scheduling algorithm for optical burst switching DWDM networks, *Proc., IEEE Globecom '01*, pp. 86–91, Dec. 2001.
7. A. Ge, F. Callegati, and L. S. Tamil, On optical burst switching and self-similar traffic, *IEEE Commun. Letters*, **4**, no. 3, pp. 98-100, Mar. 2000.
8. J. Wei and R. McFarland, Just-in-time signaling for WDM optical burst switching networks, *IEEE/OSA Journal of Lightwave Technology*, **18**, no. 12, pp. 2019–2037, Dec. 2000.
9. M. Yoo, C. Qiao, and S. Dixit, Optical burst switching for service differentiation in the next-generation optical internet, *IEEE, Commun. Mag.*, **39**, no. 2, pp. 98–104, Feb. 2001.
10. I. Baldine, G. N. Rouskas, H. G. Perros, and D. Stevenson, JumpStart: a just-in-time signaling architecture for WDM burst-switched networks, *IEEE Commun. Mag.*, **40**, no. 2, pp. 82–89, Feb. 2002.

11. K. Dolzer, C. Gauger, *et al*, Evaluation of reservation mechanisms for optical burst switching, *Int. J. Electron. Commun.*, **55**, no. 1, pp. 18–26, 2001.
12. F. Kuipers, P. Van Mieghem, *et al*, An overview of constraint-based path selection algorithms for QoS routing, *IEEE Commun.Mag.*, **40**, no.12, pp. 50–55, Dec. 2002.
13. B. Mukherjee, *Optical WDM Networks*, Springer, Dec. 2005.
14. A. Huang *et al*, Optical self-similar cluster switching (OSCS) - a novel optical switching scheme by detecting self-similar traffic, *Photonic Network Communication*, **10**, no. 3, pp. 297–308, Nov. 2005.
15. A. Huang, L. Xie, Z. Li, and A. Xu, Time-space label switching protocol (TSL-SP) - a new paradigm of network resource assignment, *Photonic Network Communication*, **6**, no. 2, pp. 169–178, Sept. 2003.

10

Hybrid Open Hash Tables for Network Processors

Dale Parson[1], Qing Ye[2], and Liang Cheng[2]

[1] Agere System, Allentown, PA 18019
[2] Laboratory Of Networking Group (LONGLAB), Computer Science and Engineering Department, Lehigh University, Bethlehem, PA 18015

10.1 Introduction

Key-based access to information is essential for many information processing systems including real-time, embedded systems. The identifying fields of application data can be taken as keys for the purpose of efficiently storing, retrieving and updating the associated information. For example, the social security number is widely used to index a person in many systems dealing with civil affairs. To reduce the length of keys, algorithms that are able to shuffle and combine the bits of keys are frequently utilized. These algorithms can turn a variable-sized amount of text into a fixed-sized output and are denoted as *hash algorithms* or *hash functions*. The reordering and reorganizing process of bits is then called as *hashing procedure*. By applying hash algorithms, a key and its associated data are turned into *(key, value)* pair and stored into a *hash table*, which is among the most common data structures used for fast information retrieval. In many embedded systems, key manipulation and access to data must occur within a hard real-time limit.

Network Processors (NPs) are special-purpose programmable embedded systems with optimized architectural features that aim to perform real-time packet-processing functions in a fast manner [1]. In general, NPs are hybrid hardware–software designs which take advantage of both the quick processing speed of hardware and the flexible programmability of software. They are expected to become the core of the fourth generation of network devices such as routers, voice over IP (VoIP) bridges and virtual private network (VPN) gateways. For example, the APP550 network processor from Agere Systems is designed to process 5 Gigabits per second [2].

The basic data flow inside a network processor is shown in Figure 10.1. The solid lines indicate direction of flow for time constrained actions. Application requirements dictate that the network processor must respond to incoming traffic within well-defined temporal bounds. Temporal constraints arise because input events may be lost or over-written if not captured with predictable speed needed to meet the throughput requirements. The time-constrained actions are also denoted as *data plane tasks*, which require a large amount of processing power. The dashed lines indicate direction of flow for processing that is not bound by temporal constraints.

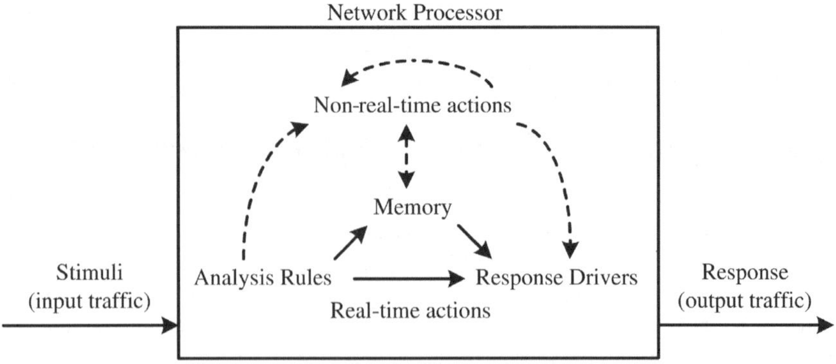

Figure 10.1. The basic data flow inside a network processor

This processing may include sampling and monitoring, statistical accounting, and processing of exception conditions. It differs from real-time processing in that it may include arbitrary iteration, using an arbitrary amount of memory in order to accomplish tasks. Real-time processing with its predictable latency constraints cannot afford unbounded iteration, and unbounded iteration comes with unbounded memory access. Real-time actions may deposit or consume data in memory, but this memory typically consists of constant-time access structures such as registers, queues, or stacks, usually implemented in hardware. The non-real-time actions are also denoted as *control plane tasks*, which require complex implementations.

To meet the requirements of properly handling a large amount of incoming traffic with high packet rates, key-based hashing algorithms are often used in the implementations of functions in NPs. For instance, *packet classification* is one of the essential tasks of network processors. It needs to cross-check multiple fields in packet headers and is required by almost all the modules along the ingress and egress paths in a network processor. For example, deciding if a packet is destined for a web server requires checking the *ETHERNET TYPE* field in its MAC layer header, the *IP TYPE* field in the network layer header, and the *TCP DESTINATION PORT* field in the transport layer header. The job of packet classification has to be done at line speed to decrease processing delay. Hash tables are useful to help the processor quickly retrieve the desired header information, by taking the combinations of the identifying fields as the key.

Another example is the *Network Address Translation (NAT)* [3, 4] function in NPs. A NAT device is usually used to connect a private IP network to the public Internet as depicted in Figure 10.2. NAT is a mechanism for translating the addresses for computers inside a private IP network into addresses that are valid in the public Internet, and vice versa. NAT service is performed for each session of communications across the border of private networks. Using the scenario in Figure 10.2 as an example, the source address of a HTTP request to CNN from 192.168.1.5 will be translated into 128.180.243.96. This address manipulation must be done at line speed

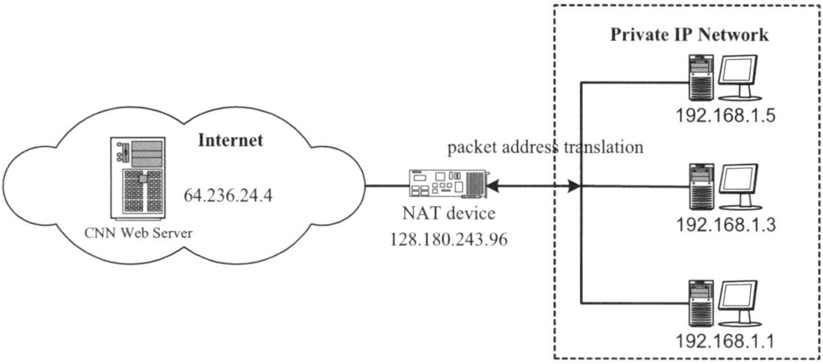

Figure 10.2. A simple example of NAT service

by taking the IP addresses as hashed keys. How to quickly and correctly map an IP address among the 4 Giga (2^{32}) address space of IPv4 is a challenge for network processor designs.

Besides NPs, hash algorithms have been proven to be effective and efficient approaches to design other network systems. In peer-to-peer (P2P) networks, resource distribution between peers in Chord [5], Pastry [6], and Tapestry [7] is performed based on a specially designed distributed hash table (DHT). Secure Hashing Algorithms (SHAs) are quite accepted for implementing network security [8] for both wired and wireless networks. Moreover, to identify a connection in the Public Internet a combination of 16-bit port number with 32-bit IPv4 address from both the source and the destination is widely taken as the hashed key.

In this chapter, we introduce the idea of hybrid open hash table designed to meet the processing time constraints of network processors. It is a composite algorithm that combines an open addressing hash table with the temporal responsiveness of incremental garbage collection. Basically, it is performed by dynamically switching between a fresh table with incremental construction via selectively copying used entries, and an aged table with incremental cleaning by emptying deleted entries. Simulation results illustrated later show that the hybrid open hash algorithm performs better than Brutil [9], a proposed chained hash table designed for real-time embedded systems.

10.2 Conventional Hash Algorithms

The study of hash tables has been a research topic for more than thirty years. Nowadays, it is one of the most common data structures implemented in many system-level libraries such as Java, STL, etc. In general, a hash table is typically constructed by mapping object keys to a relatively smaller space of buckets, which index the table. A *(key,value)* pair is called as an entry in a hash table. A bucket may contain one

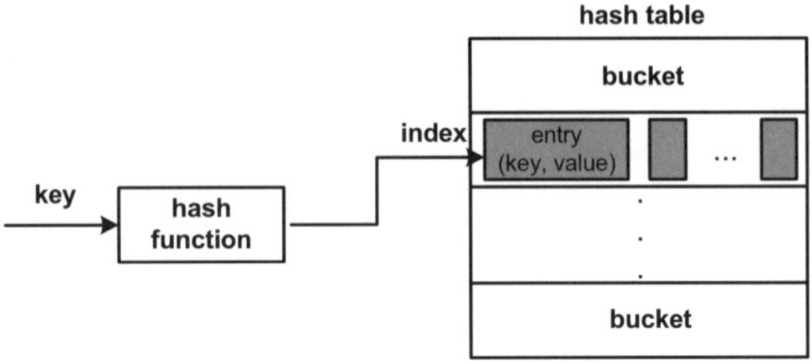

Figure 10.3. Simple structure of a hash table

or more entries based on different designs of hash table. A simple structure of hash table is illustrated in Figure 10.3.

The purpose of hashing is to generate a hashed key so as to locate the index to store the associated data. A hash function h maps every object x to a small integer $h(x)$, then stores x in bucket $h(x)$. However, *hash collision* may happen when two keys are indexed to the same bucket by a hash function. The reason is that the size of the bucket is ususally much less than the number of possible objects. Based on the structure of buckets and the strategies of handling the issue of collisions, conventional hash tables can be categorized as chained or open addressing mechanisms [10]. Both methods are able to effectivly store *(key, value)* pairs into hash tables. However, both suffer from great performance degradation over time that limits their usefulness in real-time systems that have hard timing constraints.

10.2.1 Chained Hash Tables

Chained hash algorithms map a key to a bucket, which contains a linked list of entries, *i.e.* each bucket may contain several objects hashed to the same index. Each entry in the linked list is constructed with the object key, the associated data, and a pointer to the next entry in the same chain. When hash collision occurs, a new entry that has a fresh *(key,value)* pair is added at the end the link list and the pointer of the last entry is updated. Figure 10.4 depicts the basic structure of a chained hash table.

Searching for an object takes two steps: a single hashing step to a bucket followed by a linear-time serial search through the linked list. Searching terminates when the object is found or the linked list search is exhausted. Ideally, with a good hashing function to distribute the keys uniformly into the table, each bucket's list stays short, and the search for a specific data item is quick. In the worst case, every object would be hashed to the same bucket and a long link list would be generated in just one bucket in the table. In this case, the search may have been through all the entries. This gives an unacceptable delay in real-time embedded systems in which there may be a huge number of objects.

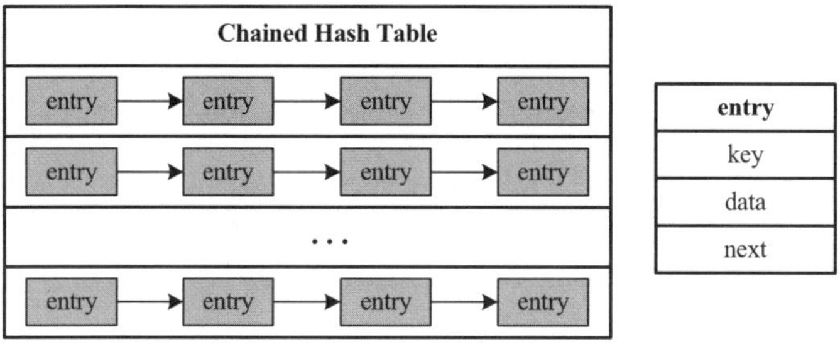

Figure 10.4. The basic structure of a chained hash table

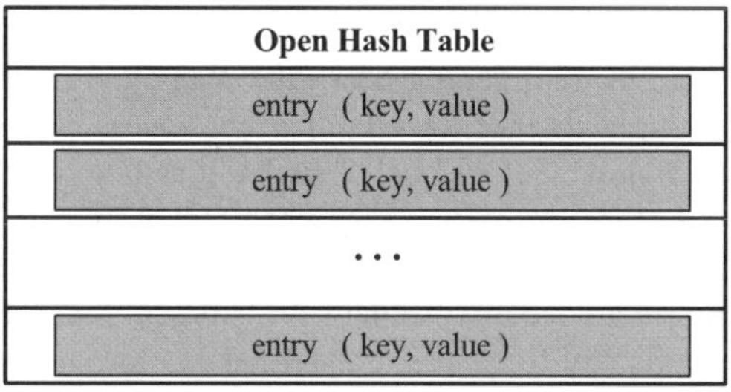

Figure 10.5. The basic structure of an open hash table

10.2.2 Open Hash Tables

Unlike chained hash tables, an open hash table maintains at most one entry in each bucket. Figure 10.5 presents the basic structure of open hash tables. When hash collision occurs (*i.e.* there is already a *(key, value)* pair in the current index), the new key that is hashed to the same bucket calls a *rehash function* to find a new bucket. Rehashing is continued until an empty entry is found. Rather than building up another data structure inside a bucket, the open addressing mechanism utilizes a list of hash functions $\{h_1, h_2, ..., h_n\}$ so that it is guaranteed that $h_i(x)$ indexes object x to different locations. The rehash function could be:

- linear probing: $h_{i+1}(x) = h_i(x) + 1$;
- quadratic probing: $h_i(x) = h_1(x) + (i-1)^2$;
- double hashing: $h_i(x) = f_1(A(x), i) + f_2(B(x), i)$, where $A(x)$ and $B(x)$ are two different hash functions, and $f_1(\cdot)$ and $f_2(\cdot)$ are two polynomial functions, *e.g.* $h_i(x) = A(x) + i{\cdot}B(x)$.

The linear probing method is often used due to its simplicity, but it is very inefficient because linear probing rehashes to a sequence of the same locations, colliding multiple times.

Searching for an object needs a single hash step to find the start bucket followed by serial rehashing steps if the desired object is not found at the index. Without an explict link list, an open hash table inserts each object into an implicit serial list defined by the hash and rehash indices of the keys. The worst case is that all the buckets indexed by rehashing functions are occupied. The object may then be discarded.

10.3 Performance Degradation Problem

The Issues

Comparing to open hash tables, a chained hash mechanism suffers the problem of large processing delays for real-time embedded systems due to the following reasons:

- *Concurrency issue.* Multithreading poses the biggest problem for chained hash tables. When multiple threads have concurrent access to a table, any of them modifying an entry must lock its bucket so that it cannot be read or written at the same time by other threads, particularly when two or more interdependent fields must be modified atomically in order to maintain consistency. This restriction is the well-known critical section problem [9]. Locking buckets will incur delays for other threads that try to access the same data. And multithreading happens quite often in network processing in order to speed up the overall performance. Open hash tables share the same need to restrict concurrency within a bucket, but the problem is worse for chained hashing because every insertion and deletion requires locking a free list of chained element objects in addition to the hash bucket's linked list) of interest, resulting in more stalls not encountered with open addressing mechanism.
- *Memory accessing delay.* For a chained hash table, the nature of the linked list data structure inside a bucket will put a colliding key into the end of the chain when collision happens. Thus, the increment of the length of the list also increases the average data accessing time because a search operation for that new key has to go through all the entries before it meets the desired one. The same scenario happens with open hashing only in the worst cases when the table is heavily loaded. Even in the zero-collision case, at least two memory access operations must occur to inspect a key residing in the table. The first access reads a hash bucket's pointer to the linked list element, and the second one reads fields from the appropriate entry. However, open hashing requires only one memory access to read a non-colliding key from a one-element bucket.

The advantages of conventional open hash tables over chained hash tables do not imply it is good enough for embedded systems that require strict predictable bounds on every system call. In fact, the conventional open addressing mechanism has a significant flaw about that it doesn't deal with deleted entries efficiently. This issue is as pernicious as the chained hash table's excessive memory access overhead.

Deleting an object from an open-addressed hash table is harder than in a chained hash table. Simply clearing out bucket $h(x)$ to delete object x could be incorrect because this bucket may be on the search path to another object which has been hashed

Bucket	Key	State
00	0005	used
01	0110	used
02	0210	used
03	0119	used
04	0099	used
05		empty

(a)

Bucket	Key	State
00		deleted
01		deleted
02		deleted
03		deleted
04	0099	used
05		empty

(b)

Figure 10.6. (a) Insert five different keys into the open hash table by hash/rehash functions. 0099 is eventually located at bucket 04 after several rehashing operations. (b) Searching for 0099 after several deletions may take a long time in this specific case.

to the same index. In the worst case, all the rehash buckets have to be examined to inspect all the possible locations of x. An exhaustive search of an open hash table is slower than the serial searching in the link list of chained addressing approach, which only needs to do hashing once.

A simple example can illustrate this performance problem very well. Suppose we are trying to map four-digit keys in the range 0000–9999 to the indices from 00 to 05 for a 6-entry hash table. In the real world, the size of hash tables used by network processors is 1000 times larger than this example. The hash function in use is to simply multiply the key's left two digits by its right two digits to form a four-digit product, from which the middle two digits are taken as the hash index. For example, the key 4567 yields a product $45 \times 67 = 3015$, with the middle digits giving a hash bucket of 01 for this hash function. Typical network processing uses a much more complicated hashing algorithm, employing bit shift and XOR operations to compress bits into a bucket. We also assume that the simplest linear probing that looks in the subsequently available bucket is taken as our rehashing function when collision occurs. Initially, the table is empty. Suppose 5 keys in the sequence 0110, 0210, 0119, 0005, and 0099 are inserted into the table with the results of hash (or rehash) indices of 1, 2, 3, 0 and 4 are used. After that, the first four keys are deleted and only key 0099 is left. The performance problem becomes obvious if we search for 0099 in the table. Conventional open hashing mechanisms will first locate at bucket 0 by hashing 0099 and find a deleted entry. Then the rehashing function of linear probing will lead the search operation to visit all the buckets from 1 to 3 till it eventually finds 0099 is in bucket 4. Figure 10.6 illustrates this case. This example shows that even when an open hash table is sparsely occupied many deletions searching may still have to go through the overall table exhaustively to locate an item, resulting in large delays that are not acceptable by network processors.

10.3.1 Improvements

Brutil [9] is an approach to improve the data accessing performance of chained hash tables designed for real-time embedded systems. It concentrates on solving the large delays that arise when a hash table becomes heavily loaded, resulting in many collisions and long linked lists in buckets. In this case, the size of the hash table needs to be extended. To do this, the conventional chained addressing mechanism will simply create a new table with more buckets (say, double the size of the original one) with another hash function whenever table extension is needed. Then it copies the overall contents from the original table and rehashes them into the new one. Obviously, these operations cause very large processing delays. Brutil's idea is to maintain two hash functions associated with two hash tables: a smaller table that is currently occupied and a pre-allocated bigger table for the future. When the smaller table becomes heavily loaded, Brutil incrementally insert entries from it into the bigger one, rather than copying the whole table at the same time. The incoming new objects will also start to be inserted into the bigger table. After all the contents of the smaller table are copied, it will be replaced by the bigger table. Brutil then pre-allocates another table with many more buckets for the next table extension. In this way, Brutil avoids the large delay between deletion of the old table and creation of the new table. However, Brutil doesn't solve all the performance degradation problems of chained hash tables, *e.g.* the concurrency issue. Our comparison results will illustrate that the use of hybrid open hashing with garbage collection outperforms Brutil in terms of total processing time.

To solve the inefficiency of deletion in the open addressing mechanism, Knuth [11] proposes an algorithm for in-place reorganization of open hashs table to remove the deleted entries that requires linear probing for collision resolution. Szymanski [12] improves Knuth's idea by removing the requirement for linear probing, but both approaches are monolithic in their table reconstruction. That means allowing table reorganization of duplicating entries within the space of one existing table. They avoid the memory cost of maintaining two tables during reconstruction, but both do not avoid the worst-case table access time encountered during reconstruction, and suffer from the same defect with respect to real-time applications. Other efforts such as Deitzfelbinger's dynamic perfect hashing [13] and Fredman's hash functions for priority queue [14] also discuss how to improve the performance of hashing algorithms, but neither of them targets real-time embedded systems.

The hybrid open hashing algorithm presented next avoids the monolithic reconstruction of tables by continually building a new hash table as it copies used entries from an aging table into the new one, skipping over deleted and empty entries. This incremental approach to reorganization is inspired by incremental copy garbage collection [15]. Garbage collection is a technique for automatically recovering storage from a running program's heap for application data structures that are no longer referenced by the running program. Classic garbage collection algorithms are monolithic — they stop all application processing during reorganization of memory, thereby impeding real-time responsiveness, similar to monolithic hash table reorganization. Real-time systems rely on incremental garbage collection, which interleaves its work with application processing in small, constant time-bound steps. Incremental copy collection achieves reorganization by copying application data into a new heap that is initially free of garbage. It is this property of incremental reorganization in the interest of avoiding monolithic stalls that the hybrid algorithm adapts to open hashing.

10.4 Hybrid Open Hash Tables

10.4.1 Basic Operations

In order to describe the hybrid hashing algorithm, it is necessary to describe three basic hashing operations: *Get* (key-based retrieval), *Put* (key-based insertion) and *Remove* (key-based deletion). In standard open hashing, Get works by searching from a key's hash index, through 0 or more rehash steps, until it finds the key; an empty entry terminates Get with failure, *i.e.* the key is not in the table. Put invokes Get to find the key; if the key is not presented, Put searches the initial hash index, rehashes used table entries, and places the key and its associated data in the first empty or deleted location found. Note that in a single-threaded architecture, Put could cache the first deleted entry location encountered by its Get call, using that location for insertion if the key is not found; this caching is not possible in a multithreaded architecture because the deleted entry may have been used for insertion of a different key by a different thread. Finally, Remove searches the hash index as Get does; if it finds the key, Remove marks that entry as deleted.

10.4.2 Basic Ideas

Two hash tables are maintained, with the *current table* being the table receiving new insertions from Put, and the *alternate table* being the aging table, from which garbage collection filters out deleted entries. Table reorganization proceeds in two phases. During the *copy phase*, both tables may contain valid keys.

In the copy phase, Get searches the current table for a key and, if it does not find the key, Get searches the alternate table; likewise Put searches both tables before inserting new entries in the current table; and Remove deletes its key from both tables. The garbage collector is invoked at the end of Get, Put and Remove to perform one table reorganization step; the garbage collector advances an index variable *cleanix* through the alternate table, one entry per invocation of the garbage collector. In the copy phase, when the garbage collector finds a used entry (*i.e.* a valid key) in the alternate table, it puts that key into the current table as a hashed insertion. Eventually, cleanix advances to the end of the alternate table, and the garbage collector moves into the *clean phase*, resetting cleanix to the start of the alternate table. At this point the garbage collector has copied all keys from the alternate table to the current table, and all new Put insertions are going into the current table.

During the clean phase, each call to the garbage collector sets all entries in one bucket within the alternate table to be empty. The alternate table is not consulted by

Get, Put or Remove during the clean phase. At the conclusion of the clean phase, when cleanix advances to the end of the alternate table, the garbage collector returns into the copy phase, resetting cleanix to the start of the alternate table. When returning to copy phase, the garbage collector reverses the roles of the current and alternate tables, so that the previous alternate table (which is now completely empty) becomes the current table (for new insertions), and the previous current table now becomes the alternate table to be filtered for deleted entries by having its used entries copied.

Table 10.1 summarizes the actions of Get, Put and Remove during the two phases of the algorithm. Incremental table reorganization could also be invoked from other functions in a system, *e.g.* by a background thread that runs during lulls in real-time activity.

Table 10.1. Operations in the hybrid open hashing algorithm

Phase	Get	Put	Remove
copy phase	Retrieve from either table.	Retrieve from either table, insert into current table if not found.	Delete from both tables.
	Garbage collector walks through alternate table, a step at a time, copying used entries into the current table via hashing. When it reaches alternate's end, it changes to the clean phase.		
clean phase	Retrieve from current table.	Retrieve from current table, insert into current table if not found.	Delete from current table.
	Garbage collector walks through alternate table, a step at a time, converting all entries to empty. When it reaches alternate's end, it changes to the copy phase, and reverses the roles of the tables (a "flip"). The new current table is empty; copying begins from the populated alternate table.		

10.4.3 Performance Evaluation

Performances Comparison of Open Hash Tables

Given the complexity that garbage collection adds to open hashing, there is a chance that the performance costs outweigh the benefits. Rather than going through complete implementations of assorted algorithm variations for performance evaluation, we compare the performances of conventional open hash tables, improved open hash tables and hybrid open tables by running representative application data through a series of related Java classes that implement an abstract Java interface *hashtablei*. This interface specifies table operations Get, Put and Remove, along with some adjunct operations and a set of measurement operations. Performance is characterized in terms of several concrete Java classes that implement hash table management strategies. Class *hashtableplain* implements standard open hashing without any table reorganization, and class *hashtablemono* adds reorganization by building a new hash table in one monolithic step when the number of deleted entries in the current table exceeds a threshold specified on the command line. *Hashtablehybrid* uses hybrid open hashing – the incremental garbage collection algorithm discussed above.

Table 10.2 shows results from a series of tests, hashing realistic Internet addresses and port numbers to hash table entries, using hashtableplain, hashtablemono, and hashtablehybrid. Hashtablemono builds a new table when a threshold of 5632 deleted entries is reached for this sample data set; the effectiveness of this threshold for the data was determined empirically. In these tests, the table can hold up to 16384 entries, with 8 entries per bucket, giving 2048 buckets. A probe count in these measurements corresponds to the number of buckets that hashing must inspect or modify to achieve one sample Put or Get or Remove operation for network traffic. Sample data was constructed so that after 8000 operations, the number of Remove operations balances the number of Put operations, eliminating the possibility of filling the table with used entries. It is trivial to see that if the Put rate consistently exceeds the Remove rate, any table must eventually run out of room.

Table 10.2. 2,000,000 operations, limit of 8000 used entries before Remove balances Put

Algorithm	Max probs	Min Probs	Average probs	Std. deviation
plain	2048	1	89.1208805	284.3068
mono	18162	1	1.746935	60.18176
hybrid	15	2	3.4318165	1.187051

The *plain* approach with no table reorganization degrades when the number of deleted entries exceeds empty entries after many deletions, because inspection of deleted entries comes to dominate the search time. The *monolithic* approach reduces the average expense of a hash table operation by periodically building a new table without deleted entries, but the worst case time of a table operation, represented here by "max probes," skyrockets because reorganization-triggering table operations must await table reorganization. Each table accessing step in table reorganization is a "probe." The *hybrid* approach of incremental table reorganization shows a 2× average probe count increase over monolithic because each hash operation incurs additional table-reorganization probes, but the worst case operation probe count drops to 15 by avoiding monolithic reorganization.

Performance Comparison of Hybrid Open Hash Table to Brutil

We conducted a comparison between the hybrid open hash algorithm and Brutil using a traffic file with 4,000,000 Internet connections. To get a fair result, both algorithms were run without multithreading and had the same initial size of hash tables. Also, a combination of IP address and port number from the traffic file was used to generate a 48-bits hashed key. The same hashing operations were performed by both approaches and the running time of Java's garbage collection was deducted for all tests during the process of simulations. We first compared the maximum execution time for running

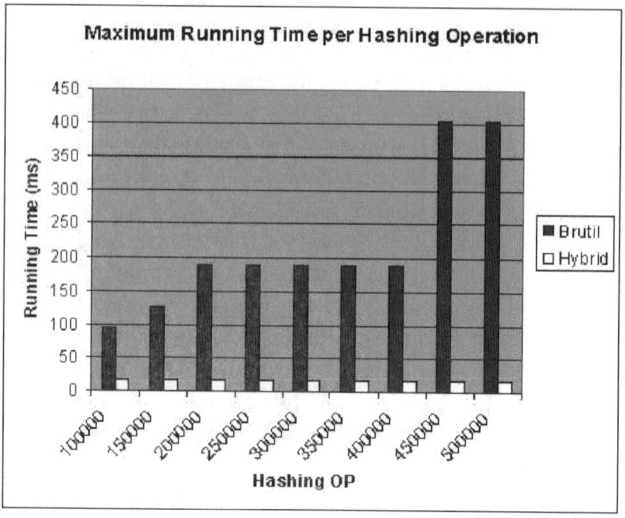

Figure 10.7. The maximum running time per hashing operation

one hash operation in these two hash tables since bounded performance is much more important for real-time embedded systems. Figure 10.7 presents the results of testing 100,000–500,000 hashing operations.

The results show that the hybrid open hashing algorithm has a bounded execution time no matter how large the number of hashing operations, while the bound on Brutil increases as the number of hashing operations increases. This means, in some worst cases Brutil may introduce very large processing delays. Figure 10.8 illustrates results for total processing time. It shows that even without multithreading, the overall performance of hybrid open hash tables is better than Brutil.

10.5 Hybrid Open Hash Table Enhancement

10.5.1 Flaws of Hybrid Open Hash Table

Given the complicated *copy-clean-switch* process of adding incremental garbage collection into open hash tables, performance costs still possibly outweigh benefits. There are two major costs in our algorithm: memory costs of pre-allocating two large hash tables, and execution time penalties by calling garbage collection too often. The former cost is not a big issue with real network processors since putting enough memory into the system can solve this issue easily. Moreover, in the initial period, the amount of consumed memory could be determined by the sum of the size of the current and alternate tables. The performance cost of the second issue could be improved by controlling the frequency of garbage collection operations.

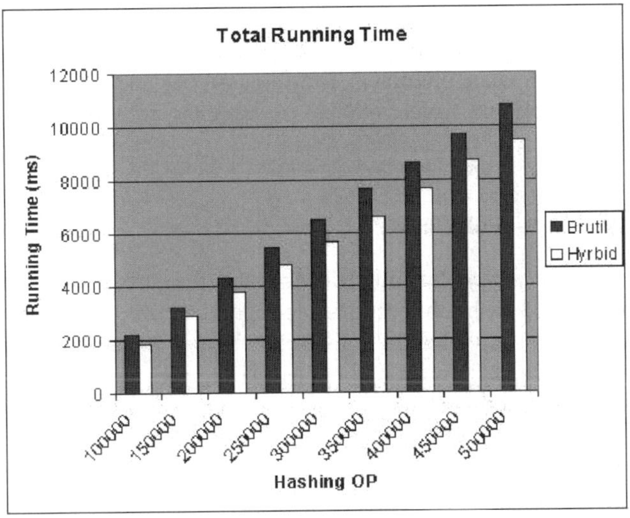

Figure 10.8. The total processing time of 10,000 to 50,000 hash operations

10.5.2 Dynamic Enhancement

In *dynamic enhancement*, an invocation of Get, Put or Remove invokes the garbage collector only when the application workload of that Get, Put or Remove has not exceeded some threshold set by the programmer. Rather than using a fixed fraction to determine how often to invoke the garbage collector, as in the basic algorithm, dynamic enhancement uses a fixed threshold. It is dynamic because it determines, on the basis of the hash table cost of each individual call to Get, Put or Remove, whether to tax that call with the additional call to the garbage collector. The main effect is to reduce the maximum number of table probes and the deviation from average required by Get, Put or Remove, because Get, Put or Remove invocations with relatively high table probe counts after doing their application work are not taxed with garbage collector calls; only Get, Put or Remove invocations with low application probe counts are taxed with table reorganization.

One limitation of dynamic enhancement is that the programmer must set the threshold, but the threshold depends on the keys and the sequence of Get, Put and Remove operations being processed. Experiments show that operations with a good percentage of Get operations work most efficiently with a threshold of one — the minimal number of probes required by a Get call that finds its key on the initial hash is one. Only those Get calls with ideal hashing need to invoke the garbage collector. Unfortunately, a long stream of operations consisting solely of Put operations with distinct keys cannot use a threshold of one, because during the copy phase, a Put operation with a new key requires a minimum of three table probes, even with no rehashing. Put must search the current table once and the alternate table once before inserting its key in the current table. The most efficient threshold for such sample sequences turns out to be three — the minimum number of probes by each Put operation. And thus any threshold less than three results in no invocations of the garbage collector, and performance plummets. But three probes add unnecessary overheads to operations containing a typical number of Get calls, where a threshold of one is the best fit.

10.5.3 Adaptative Enhancement

In *adaptive enhancement* to dynamic enhancement, the garbage collector records the minimum number of hash table probes (not counting table reorganization probes) for the minimal-cost Get, Put and Remove operation within a window of some fixed number of Get, Put and Remove operations. At the conclusion of the window, the garbage collector sets its threshold to that minimum, discarding its previous threshold. It then sets about determining a new minimum for a new sequence of Get, Put and Remove operations of the window size, repeating the process. Using a window size determined experimentally, this approach converges rapidly on an efficient table reorganization threshold for its current sample mix (*e.g.* packet traffic), yet it adapts readily to changes in the sample mix (*e.g.* traffic patterns).

10.5.4 Timeout Enhancement

A final enhancement is *timeout enhancement,* which applies to hash tables where each entry used is valid for only a finite time period after the most recent Put or Get operation on that entry's key, such as NAPT. Such tables do not provide the Remove operation; the logical equivalent of Remove occurs when an entry occupies a table location after its timeout period has expired. Each Put or Get operation that locates its keyed entry must update a timestamp field in that entry with the new expiration time for that key. Searches that detect entries with expired timestamps treat those entries as having deleted status; garbage collection does not copy these deleted entries. The advantage of this approach over alternatives such as content addressable memory (CAM) is that it is not necessary to use interrupt timers or other active means to search the hash table to remove expired entries. By treating expired hash table entries as deleted entries during normal search, timeout enhancement avoids explicit searches for expired entries. CAM, on the other hand, appears to require explicit timeout instruction processing for expired entries, because there is no "normal search" that can convert expired used entries to deleted entries.

10.5.5 Performance Evaluation

Table 10.3 shows measurements for some of these enhancements of the basic hybrid open hash algorithm under the same scenario as that of Table 10.2. The *hybrid* $\frac{1}{2}$, shows slight improvements if the garbage collector (GC) is invoked for only one half of the Get, Put or Remove operations instead of every operation. Timeout replaces Remove with expiration-based timeouts and invokes GC on every Put and Get call;

Table 10.3. 2,000,000 operations, limit of 8000 used entries before Remove balances Put

Algorithm	Max probs	Min probs	Average probs	Std. deviation
hybrid $\frac{1}{2}$	14	1	2.6059785	1.394060
timeout	8	2	3.2463965	1.032094
timeout $\frac{1}{2}$	8	1	2.475426	1.130879
dynamic 3-4	6	2	3.245097	1.030479
dynamic 1-2	6	2	2.496241	0.5020706
adaptive	6	2	2.496241	0.5020706

timeout treats Remove calls in the sample data set as Get calls. *Timeout* $\frac{1}{2}$ invokes the GC on half the calls. Reductions occur because two-table deletion searches have been eliminated. *Dynamic 3-4* uses timeouts and it invokes GC with a threshold of 3 during the copy phase and 4 during the clean phase – the clean phase of GC is less expensive, allowing a higher threshold – while *dynamic 1-2* uses thresholds of 1 and 2 respectively. Only Put and Get operations whose probes are less than or equal to the threshold invoke GC. *Adaptive* uses adaptive enhancement with a window size of 64 operations to set the thresholds to the minimum probe counts during those windows. For normal traffic it is identical to dynamic 1-2.

10.6 Extended Discussions of Concurrency Issues

There are some concurrency issues to consider in the two-table approach of the hybrid table reorganization algorithm and its enhancements. Bucket locking of multithreading does not solve all problems.

10.6.1 Insertion

The first issue has to do with concurrent attempts to insert an identical key within the current table. Suppose Thread_A and Thread_B both attempt to insert new key K into the current table. For argument's sake suppose that garbage collection is in the clean phase, so that the alternate table is not consulted before insertion. Both threads attempt to find K using Get but fail because it is not in the table. Both must restart searching the initial hash index. As previously mentioned, they cannot save the location of the first deleted entry because it may subsequently have been used by another insertion.

Suppose Thread_A re-enters the current table first and finds entry E_X to be occupied. Thread_A continues searching for a deleted or empty entry. Suppose further that, before Thread_B arrives at E_X, the timestamp in E_X expires, so that Thread_B considers E_X to be deleted. Thread_B now inserts K at E_X while Thread_A inserts K at a subsequent entry in the search path. Key K now erroneously appears redundantly in the table.

The solution is to test for entries that are about to expire within a very small time within the Put portion of insertion, after the attempt at Get has failed, and increase their life by a small amount. The increment must be significantly smaller than the normal decay periods for UDP or TCP mappings. This PayloadPlus implementation adds 1 second to the expiration deadline of any occupied entry along a Put search path that is about to expire in less than 1 second. This 1 second "refresh" is more than enough to avoid the problem, but it is small enough in comparison to normal decay rates so that, as long as hashing and rehashing scatter searching enough to avoid consistently refreshing the same entries, it does not increase overall longevity of expired mappings significantly. In PayloadPlus this test occurs entirely within an ASL Put function running within the policing engine.

10.6.2 Clean-to-copy Phase Change

Another concurrency issue arises when the phase of garbage collection changes. When changing from clean to copy phase, the current table and the alternate table change places. Suppose Thread_A begins to store key K into table T_0 (as the current table), then a clean-to-copy "flip" occurs, making T_1 (the current table), and then Thread_B begins to store K into T_1. The result is redundant insertion of K into both tables.

The solution is a so-called *write barrier* [15]. A barrier is constructed so that active threads for Put operations stall a table flip until their Put operations are completed; and a pending table flip stalls any new attempts to conduct Put operations until it performs the flip. Table mutation becomes a pipeline of grouped Put operations and single flip operations. In the example, suppose a flip arrives between Thread_A's and Thread_B's attempts to insert K. Thread_A completes its insertion, then the flip occurs, then Thread_B begins the initial Get portion of insertion. Since Thread_A has completed insertion, Thread_B will find K. If Thread_A and Thread_B had both attempted to perform Put without separation by a flip, they would have attempted insertion in parallel, and the first thread that had arrived at a deleted or empty entry in the current table would have inserted the key; and the other would then have found it.

The write barrier is also necessary to avoid the flip while there are still threads performing clean phase garbage collection. If a flip occurs while there is still cleaning of a table, an insertion into that table could be undone by an immediate cleaning of its bucket. Cleaning threads are treated like Putting threads with respect to the write barrier. There could also be a critical section problem with the change from the copy to the clean phase. Even though this change does not flip the current and alternate tables, it could hide the alternate table from Get searches before earlier garbage collecting threads have completed copying it; the alternate table is not searched by Get during clean phase, but if not all entries have been copied before the phase change occurs, some entries would "disappear" momentarily until their copying completes.

The write barrier for the clean-to-copy flip entails multithreaded synchronization over two counters for threads, which are performing Put and awaiting Put, and a flag bit for the pending flip. This synchronization can only stall threads at the time of a flip. Fortunately this condition arises very infrequently in the garbage collection cycle. Waiting for a flip is not a major source of delay for Putting threads.

10.6.3 Timestamps

A final, easily handled critical section issue is the matter of copying timestamps during the copy phase of garbage collection. An entry copied from the alternate to the current table must have its timestamp copied, but not refreshed, to retain its application-driven decay deadline. But, it is difficult to copy a 48-bit key and a 16-bit timestamp simultaneously in one function invocation due to the hardware limitation of network processors. It requires two, violating atomicity. Between these two calls another application thread could refresh the key by using it, in which case the old timestamp should not be copied because it is now outdated. The solution is to have a policing function that copies in the key set a minimal timestamp — again a decay time of 1 second — and have the second function that copies in the timestamp test whether its copied timestamp would decrease the life of the mapping. Normally it will not, but if an intervening thread has refreshed the timestamp, then the timestamp-copying policing function avoids storing its timestamp argument in the policing flow.

10.7 Conclusion

For real-time embedded systems such as network processors, conventional hash tables have the problems of performance degradation due to concurrency issues with multithreading, large memory accessing delays and not dealing with deleted entries efficiently. In this paper, we propose an approach combining hybrid open hash tables with incremental garbage collection to meet the needs of real-time applications. By maintaining two hash tables instead of one table as in conventional approaches, the hybrid open hashing approach incrementally copies the valid keys from the alternate table into the current table, skipping over the deleted or empty entries. In this way, hashing operations always deal with a table without too many hash collisions. Performance evaluations show that the hybrid open hash table is better than Brutil and it still has the potential to be improved further by several enhancement approaches.

References

1. M. Peyravian and J. Calvignac. Fundamental architectural considerations for network processors. *IEEE Journal of Computer Networks*, pp. 587-600, issue 41, 2003.
2. D. E. Comer. *Network Systems Design using Network Processors, Agere Version.* Upper Saddle River, NJ: Prentice Hall, 2004.
3. D. Comer. *Computer Networks and Internets with Internet Applications,* Third Edition. Upper Saddle River, NJ: Prentice-Hall, 2001.
4. *Traditional IP Network Address Translator (Traditional NAT).* Internet Engineering Task Force and the Internet Engineering Steering Group, Request for Comments 3022, Jan. 2001.
5. I. Stoica, Robert Morris, David Karger, M. Frans Kaashoek, and Hari Balakrishnan. Chord: a scalable peer-to-peer lookup service for Internet applications. In *Proceedings of ACM SIGCOMM 2001*, San Diego, CA, Aug. 2001.
6. A. Rowstron and P. Druschel. Pastry: Scalable, distributed object location and routing for large-scale peer-to-peer systems. IFIP/ACM International Conference on Distributed Systems Platforms (Middleware), Heidelberg, Germany, pages 329-350, Nov. 2001.
7. Ben Y. Zhao, Ling Huang, Jeremy Stribling, Sean C. Rhea, Anthony D. Joseph and John Kubiatowicz. Tapestry: A Resilient Global-scale Overlay for Service Deployment. *IEEE Journal on Selected Areas in Communications*, **22**, No. 1, Jan. 2004.
8. RFC 3174. http://www.faqs.org/rfcs/rfc3174.html, accessed on Sep. 6, 2005.
9. S. Friedman, N. Leidenfrost, B. Brodie and R. Cytron. Hashtables for embedded and real-time systems. In *Proceedings of IEEE Real-time Embedded System Workshop*, Dec. 2001.
10. T. H. Cormen, Charles E. Leiserson, and Ronald L. Rivest. *Introduction to Algorithms.* The MIT Press, 1989.
11. D. Knuth. *The Art of Computer Programming.* Vol. 3, *Searching and Sorting*, Reading, MA: Addison-Wesley, 1973.
12. T. Szymanski. Hash table reorganization. *Journal of Algorithms*, **6**(3), pp. 322-335, 1985.
13. M. Dietzfelbinger and F. Meyer auf der Hyde. An optimal parallel dictionary. In *Prococeeding of ACM Symposium on Parallel Algorithms and Architectures*, pp. 360-368, 1989.
14. M. Ajtai, M. Fredman, and J. Komlós. Hash functions for priority queues. In *24th Annual Symposium on Foundations of Computer Science*, pp. 299-303, Tucson, Arizona, 7-9 Nov. 1983.
15. R. Jones. *Garbage Collection, Algorithms for Automatic Dynamic Memory Management.* Chichester, John Wiley & Sons, 1999.

Index

A

ADJUST algorithm, 34

B

Batch switching, 66
Buffered crossbar switches, 12

C

Chained hash tables, 215
Combined input output queued (CIOQ) switch, 10, 57
Combined input-crosspoint buffered (CICB) switch, 128, 143, 170
Cone algorithms for packet scheduling, 89

D

Delay bounds, 49, 56
DOUBLE algorithm, 33
Dual scheduling algorithm, 148

F

Fabric on a chip (FoC) Field Programmable gate array (FPGA), 120, 102

G

Geometry of packet switching, 82

I

Internally buffered crossbars, 127
iSLIP algorithm, 7, 124

L

Longest queue first (LQF), 6, 133, 172
Lyapunov methodology, 46

M

Markov chains, 46, 109, 151
Maximal weight matching, 56
Maximum weight matching, 6
Micro-electro-mechanical systems (MEMS), 29
Multi-stage switching, 20

N

Network processor, 212
Networks of switches, 60

O

Oldest cell first (OCF), 6, 54, 132, 172
Open hash tables, 212
Optical burst switching (OBS), 199
Optical fabrics, 28
Optical packet switching, 27
Opto-electrical switching, 3, 27
Output queued switch emulation, 104

P

Parallel iterative matching (PIM), 7

Q

Quality of service (QoS), 94, 104
Queue size bounds, 49

R

Round-robin arbitration, 6, 99, 133, 137, 183

S

Stability definition, 45, 87, 155
Stability region of input queued switches, 53
Strong stability, 45
Switch architectures:
- combined input–output queued, 10, 42, 126
- input-queued 4, 83, 131, 148
- output-queued, 4, 104

T

Throughput optimal scheduling, 160
Time slot assignment, 31
Time space label switching, 198
Traffic models, 136, 177

V

Virtual crosspoint queueing, 190
Virtual output queueing (VOQ), 5, 84, 141, 171
Weak stability, 45
Work conservation, 12

Printing: Krips bv, Meppel
Binding: Stürtz, Würzburg